*e*volution

Also by the same author

Crooked Minds: Creating an Innovative Society

*e*volution
DECODING INDIA'S
disruptive ↑ TECH STORY

KIRAN KARNIK

RUPA

Published by
Rupa Publications India Pvt. Ltd 2018
7/16, Ansari Road, Daryaganj
New Delhi 110002

Sales Centres:
Allahabad Bengaluru Chennai
Hyderabad Jaipur Kathmandu
Kolkata Mumbai

ISBN: 978-93-5333-256-3

First impression 2018

10 9 8 7 6 5 4 2 3 1

The moral right of the author has been asserted.

CONTENTS

Author's Note / vii

Introduction / xiii

Money at Light-speed / 1

Dr Google's Clinic / 22

To Silicon Sir, with Love / 39

ABC and Technology / 60

Technology, Governance and Democracy / 82

Soft Power: Mind versus Munitions / 101

Technology of the Future, and the Future of Technology / 123

India's Tech Triumphs / 141

Epilogue / 166

Endnotes / 179

Index / 191

AUTHOR'S NOTE

The interplay of technology and daily life has long fascinated me. This is probably the result of not being an engineer and yet being deeply involved with technology for almost all of the last five decades. Technology has moved from the vacuum tube (how many today have even heard of a vacuum tube, leave alone knowing what it is?) to microelectronics, with one small chip of the latter having more capability than a room full of the former. Electric cars may soon replace conventional ones, while fully automated smart cars may displace drivers. Mobile phones have taken over from fixed lines or landlines. Holidays in Goa or Bali may soon be replaced by ones in outer space, and invitations to friends' hundredth birthdays may become routine.

The changes in the homes of the middle class in India are also striking. I have witnessed the transition from wick-burning kerosene stoves to piped gas and microwave ovens. Dishwashers and washing machines are now common, as are vacuum cleaners. Fans have not gone away, but air-conditioners and room heaters abound. The radio has yielded its pride of place in living rooms to the satellite-linked television, though the young would rather watch shows by an online streaming service on their laptops. The cycle in the driveway has made way for the scooter, the motorcycle, or even a car. Terraces that saw Yagi antennas for TV, made way for the satellite dish, and may soon serve as parking areas for personal transportation drones.

All this tells us the story of not only the recent history and evolution of technology, but also of the rapidly changing and evolving socio-economic scenario in the country. It is the confluence of these two that is at the heart of this book.

I have chosen to look at how technology is influencing and affecting some key elements of daily life: money and finance, health, education, attitudes and behaviour, and accessing government services. I have touched upon an aspect that intrigues me: the interaction of culture and technology, and how one affects the other. I have picked simple, day-to-day examples—most of which have been personally experienced. This is obviously set in a broader context, and so the book also looks at technology and democracy, at how soft power is evolving with the use of technology, and where technology may be headed in the near future. I have also touched upon areas I feel concerned about: surveillance (ostensibly to protect our freedom) and cybersecurity.

Having spent so many years in organizations handling nuclear, space, and information and communications technologies, I could not resist the temptation of extracting learnings from these areas of India's success, and using them to make recommendations for strategies in emerging technologies. These suggestions will, I hope, stir dialogue and discussion.

Therefore, as mentioned, the book is substantially based on personal experience and what I have seen, rather than being a purely academic or research-based treatise on technology and its impact. It gives expression to many of my own thoughts, concerns and hopes. I am grateful to Rupa Publications for the opportunity to share these with a wider audience. This could not have happened without the personal interest and deep involvement of Kapish Mehra, Managing Director of Rupa. I am much impressed by not only the time that he spends with authors, but by his considerable knowledge about information technology (IT) and IT-enabled services. His understanding and insights often made me wonder whether he was secretly working within the IT industry!

My manuscript greatly benefited from the careful editing and corrections by Yamini Chowdhury, Senior Commissioning Editor in Rupa. She worked with me from the conception of the book, to its final stage, occasionally reminding me of missed deadlines—but always gently and very politely! More importantly, her comments, questions and suggestions have helped sharpen many passages in this book. I have deep appreciation for her inputs and efforts. Of course, she let me have the last word, so any failings that you might see are completely my responsibility.

Besides the above, I would also like to thank Debangana Banerjee, Assistant Copy Editor at Rupa, for painstakingly editing the manuscript.

I must acknowledge, with gratitude, the contribution of a number of people who helped me with specific inputs, or corrected particular paragraphs with regard to facts.

Since I do not operate from any fixed office, practically all the writing was done at home. This meant possible interruptions of one kind or another and a host of distractions. It also made procrastination and laziness easy! Protecting me against these, and ensuring that I stay on the job, was the task taken on by the mistress of the house, Sunitee. There was no gentle or polite prodding here, though—only direct assaults to ensure that writing the book stayed a priority on my agenda. On the occasions when my motivation was high and the flow of words fluent, she often put up with long delays for lunch, or for departures from planned outings, while I wrote. Her support and unfailing encouragement have ensured that this book sees the light of day.

Finally, a word about the biggest contributor to this book: she's the one who played multiple roles—the occasional critic, pointing out the repetition of ideas or words, and spotting any fuzzy thinking or gaps in the narrative; the frequent researcher, helping me to quickly verify a name, fact, date, or an article on a particular subject; and finally, the stenographer and typist, who pitied my slow and sloppy efforts, and demonstrated the speed at which a more efficient person could do the job. She helped to organize the files and integrate various pieces appropriately. Some of this was done through collaborative work via long-distance conference calls (an immediate and very practical example of information and communication technology [ICT] that I have written about). And all this was done by burning the midnight oil and during her off-time from a full-time job. Without the help of my daughter, Ketaki, this book would have been far less than what it is, and would have still been at the draft stage.

I must also thank those of you who have reached so far. Do read the rest of the book!

INTRODUCTION

Navigating the Uncertain Waters of the Future

From the beginning of history, technology has been the greatest force in shaping the evolution of human society and driving its economic progress. Centuries ago, the discovery of fire, and the technology to create it, resulted in radical changes in how humans lived. Today, new technologies emerging from the 'electronics revolution' are causing similar monumental changes in the social and economic realm. In addition, advances in fields like biology, additive manufacturing (3D printing), artificial intelligence (AI) and machine learning are catalysing major transformations. The combination of these technological advances in various areas, and the resulting synergy, have brought about the so-called 'fourth industrial revolution'. While the name and focus is on industrial impact, the societal changes could well be far more important and profound.

New technologies are, in fact, bringing about change faster than ever, with technology itself driving its progress. Take the example of ICT. As new ideas are born and new technologies are created in this field, they get communicated around the world faster, thanks to ICT. This enables others to quickly build on them further, thus creating a continuous cycle of accelerating progress in ICT and its applications. Further, ICT facilitates online and real-time collaboration amongst people who are thousands of kilometres apart. This enables the creation of global teams comprising the best brains in a field, irrespective of their location—thus adding to the extent of new technological breakthroughs.

The rapid development of some of these technologies is now giving rise to serious concerns about the dangers of a new man–machine equation—one in which machines may not only displace, but dominate man. That such a scenario is no longer a wild, dystopian science-fiction story is borne out by the credentials of those who consider this a serious danger. Eminent scientist, the late Stephen Hawking, the co-founder of Apple, Steve Wozniak, and hundreds of others wrote an open letter in 2015, warning that AI can potentially be more dangerous than nuclear weapons.[1] Elon Musk, founder of Tesla and SpaceX, has characterized it as 'our greatest existential threat'.

Though this species-threatening possibility is, till now, only in the realm of conjecture, there are many major changes that we have already witnessed over the last few decades in diverse areas of human endeavour, which are all driven mainly by technology. It has, for example, completely changed the world of banking and finance, both at the level of organizations as well as individuals. Healthcare and medicine have already been transformed and the day is not far when it will be routine for people to celebrate their hundredth birthdays. In the area of education, it is still early to ascertain the impact of technology, but the near future is slated to see extraordinary changes, including the likely demise of 'chalk and talk' pedagogy. Last but not the least, thanks to technology, trade and globalization are locked in a mutually reinforcing upward spiral.

Even as technology-driven changes in these areas begin to impact us in myriad ways, the most substantial change may well be in our attitudes and values. This is obviously a slower and more subtle process, but there's no doubt that it is under way. Meanwhile, the quick and easy flow of ideas across the globe—driven by powerful communication technologies that are accessible at ever lower costs—has begun to tilt the balance away from sheer military strength as a definition of geopolitical power.

There is little doubt that technology has brought economic well-being and a more comfortable life to much of humanity. At the same time, there are newly emerging concerns. One such concern is climate change. This poses a serious threat to not only the economic fortune of billions, but threatens the very survival of a vast number of people. A rise in sea levels caused by global warming is likely to result in the inundation of large tracts of the coastline in many countries and the submergence of islands. Climate change will not only make history, it will also alter geography. A few countries, mainly low-lying island nations, may well disappear from the map. One can hold technology responsible for this, as global warming is the direct consequence of pollution caused mainly by industrialization. Yet, it may well be that the villain of the piece turns out to be the saviour: it is technology that can help mitigate—and possibly reverse—the effects of pollution.

A BETTER UNDERSTANDING OF 'OUR GREATEST EXISTENTIAL THREAT'

The fourth industrial revolution—unlike past ones—promises economic growth, with few of the side-effects seen earlier. Technologies like electric or hydrogen-powered vehicles, solar and other forms of renewable energy, smart grids and intelligent homes, which help lower power consumption, will contribute to halting and possibly even reversing global warming.

Technology may be the saviour in a different way, too. If global warming cannot even be halted, Earth may well become largely uninhabitable. This could also happen in the case of a large-scale nuclear war, or in the very unlikely, but not impossible, event of a cataclysmic asteroid collision. All these point to the desirability of having an alternative 'safe haven'. Technology could provide this in the form of the development of one or more extraterrestrial settlements.

Understanding how technology has affected us in the past, and is today influencing our lives even more, is essential if we want to chart a course through the uncertain waters of the future. To some, accelerating technological change and new technologies might seem to pose a threat. However, historically, new technologies—even those that are disruptive of established arrangements—have always created exciting new opportunities.

There is no reason why the technologies of the fourth industrial revolution will not do so too.

In keeping with this, the attempt here is to first identify and understand the changes that technology has wrought in various fields, over the last few decades. I, then, seek to indicate the likely impact of technology in these areas in the coming years, with the full knowledge that any such predictions from 'crystal-ball gazing' are hazardous when technology is advancing so fast. Specifically, in the next three chapters, I will look at technology's interplay with the world of money and finance, health, and education—first, comparing the past and present, and then discussing what the future may hold. Thereafter, I will explore the changes that technology has brought about in the broader realms of attitude, behaviour and culture (ABC). I will also examine the obverse: how ABC has influenced the ways in which technology is actually used. I will, then, move on to describe the impact that technology has had on governance (at the level of the individual's interaction with the government, as well as overall governance) and on various facets of democracy. Shifting gears and going beyond the direct impact of technology on an individual, the next chapter will look at soft power, focussing on the role of technology in the worldwide battle for winning hearts and minds. The narrative will, then, move on to what may be yet to come: the technologies of the future, and the future of technology. Finally, I will analyse India's technological success and suggest how India might—through concerted action and within a clearly-thought-out policy framework—become a leader in emerging technologies, so as to take full advantage of all the possibilities and benefit from these.

Many of the technology-driven changes that we have witnessed are highlighted through examples and anecdotes based on my own experiences, drawing on a personal association with premier organizations in the areas of nuclear energy, space, broadcasting and ICT. These examples seek to provide a practical and contextualized understanding of the sweeping changes brought about in key areas, thanks to technology. The objective is not only to document the impact of technology on the sectors under study, but also to look at future possibilities and foresee what effects they may have.

Ultimately, my aim is to generate discussion, spur dissenting views and trigger further analysis—all originating from different perspectives—thereby helping our country to not only cope with the emerging future, but also to try shaping it to the benefit of all.

1

MONEY AT LIGHT-SPEED
Technology Makes Money Go Round

'Money makes the world go round'—this seems an obvious truism in the modern economy, where finance plays such a predominant role. Little wonder, then, that one of the fastest and biggest users of technology is the world of money and finance. As society moved from barter to commonly accepted forms of 'stored value' (gold, silver and coins), trade became easier. One could now sell goods or services to one customer and buy what one needed from someone else, using some form of stored value. This could also be used to accumulate wealth, which could then be used to buy whatever one wanted, whenever one desired.

Currency notes are the more modern form of stored value. However, sending these over a distance takes a lot of time and effort, especially if it is across countries. At the same time, growth in global trade has made it necessary to move funds from one country to another on a scale that is many times what it was till only a few years ago. Apart from the organization-to-organization transfer of funds that this involves, there are also payments across borders made by, and to, individuals. This movement of money is obviously not a physical transportation; it is only the information that is sent—about the amount, currency, account numbers and other details—which helps to authenticate the transaction. All this information used to be sent through messengers, mail or telegram; now it is sent over the Internet, or via dedicated communication links.

ACUTE WITHDRAWAL SYMPTOMS

The speed, reliability and authenticity of such cross-border transactions oil the wheels of global trade, facilitating the movement of goods and services. Today, an individual can surf an e-commerce website to find and order goods from elsewhere in the world. The whole transaction, including cross-border payment by credit card or bank transfer, is done from the comfort of one's home. This is now routine and a simple process for the consumer, but behind this lie a host of technologies related to communication and connectivity, the Internet, security, verification protocols, and software and mobile applications. This is now part of a broader trend: to shift all complexity to the back-end and make the user's experience as easy and comfortable as possible. Thus, it is not surprising that there are mobile phone apps for complex tasks, which can be used even by a semi-literate, 'non-tech' person. One example is payment apps. Now, a payment can be effected from one's mobile phone merely by scanning the Quick Response (QR) code of the seller and typing in the amount. This is being used across India, including by those who are illiterate. What a radical change in just a few years!

It was not too long ago that you could access the money in your own bank account only between usually 10 a.m. and 2 p.m., and that, too, on weekdays and, as a special concession, on alternate Saturdays (excluding holidays). Thus, you had no access to your money for about a hundred days a year and for over 80 per cent of the day. Further, you could withdraw money only from the city, bank and branch in which you had an account.

As a result, cash withdrawal required much planning: making sure it was not a holiday; taking time off from your work (or other activities) in the six- or seven-hour window when the bank was open; writing out a cheque or withdrawal slip; travelling to your bank branch; waiting in the inevitable queue at the withdrawal counter; and praying that when you finally reach the front of the line, the functionary concerned does not take an informal—and uncertainly long—tea, lunch or toilet break. After all these steps, you would get a 'token' in exchange for your cheque or cash withdrawal slip. With this in your pocket, you then typically stood around (there being no place to sit) somewhere near the cash counter, along with many others, and waited for your turn. In old branches, someone behind the counter would yell out the number; in 'modern' ones, a display would show the token number being handled next.

You could, while you waited with one eye on the display, try and get your passbook updated. This meant another queue and, when your turn came, handing over your passbook for the updation of the record of transactions in your account. The entries were, of course, made manually—a task for which the clerk took his own sweet time. More often than not, you would be told to leave the passbook for updation and pick it up the following day.

By the time your passbook got updated, hopefully your token number would have come up, enabling you to go to the cash counter and collect your money. A few hours would well have gone by, meaning that you had to take a half-day leave from work, just for withdrawing money. If one monetized this and added the cost of travelling to and from your bank branch, the amount involved would be far from trivial. This meant that you would end up spending a fair amount to withdraw your own money from your bank account.

In the '70s, I had to travel fairly frequently. While there were many differences and inconveniences in out-of-town trips then, the biggest issue was related to money. For those who, like me, travelled out of their home city, finances were required to be planned well in advance. Being in a different city meant you had no access to your bank account. The only practical option was to estimate your likely expenditure, add a safety margin and carry that amount of cash with you. If you underestimated your need (as had happened with me more than once), you would have no alternative but to depend on friends in the other city for a temporary loan. If you had no acquaintance there, you would have little choice but to economize on (or even give up!) meals. Remember that credit cards were extremely rare, so even large payments (a hotel room, for example) had to be made in cash. Even if the relationship was one of trust, no one would accept a cheque because an 'outstation' cheque would take days (generally more than a week) to be realized.

For those below forty years of age, all this may seem like tall tales or, at least, a much exaggerated account. Those above sixty, though, would vouch for the veracity of this, and narrate personal experiences of their own travails of sometimes even having to spend an entire day in the task of cash withdrawal. Of course, there was the inevitable silver lining. In this case, it was the human angle of the bank staff being familiar with their customers, since one had to go to the same bank branch regularly for all transactions. Friendships sometimes sprouted and I know of one case where regular visits and interaction between a young account holder and a counter clerk blossomed into marriage.

THE HOLE IN THE WALL THAT CHANGED OUR WORLD

After the computerization of rail-ticket reservations began in 1985, the unheralded Automated Teller Machine (ATM) revolution was the biggest and most impactful outcome of new technology in India. It required high-reliability connectivity to ATMs across the country, and much of this was provided through satellite communication links. It needed computerization in banks, with real-time updating of bank accounts, databases and strong security features. At the front end, the ATM itself had to be sturdy, safe and reliable, while providing a simple and easy interface with the customer. The fact that the system generally works smoothly, with minimal glitches and problems, is certainly a feather in the cap of the banking system. It is also a powerful example of the use of technology for the benefit of the common man.

Now, no one thinks of the time of the day, the day of the week or holidays, when they want to withdraw money. Going to the branch that hosts your account, or even to the same bank, is not a thought that crosses one's mind. Little wonder, then, that so many young people have never been to the branch where they have an account, after the first visit! All one does is stop at any ATM, walk in, insert one's card, enter one's PIN and amount, and walk out with the cash—all in a matter of minutes, at any time of the day or night, on a holiday or working day, and in any part of the country. Further, one can get a printout of their last few transactions and the balance in their account.

No one has fully quantified the hours and effort saved in the move from the old system to the ATMs, but the resultant saving in monetary terms would certainly be massive. In addition, with fewer customers going to banks for withdrawals, the latter need less space and staff, resulting in more savings. This enables them to extend their outreach more easily.

From ATMs linked only to specific banks, the nudging and policies of the regulator—the Reserve Bank of India (RBI)—have led to multiple-bank ATMs. This means that a customer can go to any bank's ATM and complete transactions for his/her account, irrespective of the bank in which they have their account, or its location. Now, there are also 'white-branded' ATMs, run by independent third-party providers and accessible to the customers of any bank. As of mid-2018, an estimated 200,000 ATMs across the country[2] facilitate easy cash withdrawals for account holders.

TIME TO GO DIGITAL

Like the earlier 'token' and human teller system, the ATMs too are now facing obsolescence, as new technologies threaten their existence. Digital money is beginning to make its way into the system and the demonetization of high-value notes (₹500 and ₹1,000) in November 2016 gave a big fillip to digital payments. The huge queues at ATMs and the limited availability of cash pushed people into adopting the mode of digital payments, especially for smaller amounts. Mobile phone-based apps, like Paytm, got immense traction as people discovered their advantages in terms of ease and simplicity of operation. Even semi-literate people quickly learned how to use such payment apps.

Most people had little choice but to 'go digital' as the shortage of currency after demonetization had the inevitable effect of encouraging the hoarding of notes (and even coins!), which led to even greater shortage. Daily transactions, like buying vegetables, milk and bread, became difficult. Since point of sale (POS) machines for credit/debit cards were not all that widespread (the local vegetable vendor certainly did not have one!), the use of cards ('plastic money') was limited to bigger shops and purchases. It was in this scenario that mobile phone payment apps began to be seen as an excellent alternative. While most such apps require a smartphone, some were developed to also be able to run on cheaper feature phones. As long as both parties—buyer and seller—had the app on their phones, they could transfer money instantaneously. Preloaded money in one phone's app could be transferred through just a few easy steps, to another. The ease and user-friendliness of the process led to even roadside vendors adopting it. As a pioneer with established brand repute, Paytm benefited the most from this move to digital payments, becoming almost synonymous with payments through mobile phones. Signs like 'Paytm accepted here' became common in small shops and vendors' handcarts across the country. During the demonetization period, Paytm wallet users reportedly increased from 125 million to 185 million in three months, going up to 280 million by November 2017.[3]

The government catalysed payment mechanisms like the Bharat Interface for Money (BHIM) app. Earlier, money transfer between bank accounts was facilitated through systems like National Electronic Funds Transfer (NEFT) and Real-Time Gross Settlement (RTGS).

CASH IS STILL KING

All the above methods of electronic payment may gradually render both paper-based (cheques/bank drafts) and plastic-based (credit/debit cards) means obsolete. The move towards a 'less cash' society—articulated explicitly in the RBI's Vision 2018 document for payment systems[4]—is now actively under way.

Of course, the convenience, fungibility and anonymity of cash mean that for many people, cash is still king when it comes to monetary transactions. In addition, till the infrastructure becomes highly reliable, connectivity and speed may sometimes create difficulties for digital payments. As a result, cash is unlikely to go away any time soon. The dream of a cashless society is likely to remain just that—a dream—at least, for a long time to come. Therefore, 'less cash' is a more practical aspiration than 'cashless'.

EMERGING TRENDS AND NEW TECHNOLOGIES

A new use of the capabilities of the Internet and digital technology is for crowdfunding. This enables an entrepreneur to raise funds for his or her venture by posting details online and inviting all those interested, to commit funding. Thus, the entrepreneur is able to tap into a very large base of potential investors who visit the site and would like to fund the particular project. Casting such a wide net for funding would have been impossible without the Internet. This new way of reaching out to a very large number of people has seen some exceptional success. One example is a project by a non-profit organization, Suryoday Parivar, to build an 8-km-long water canal in the village of Horti, located in the Osmanabad district of Maharashtra, to serve 700 farmers.[5] This was a collaborative effort, with the farmers pitching in and contributing 50 per cent of the total cost of the project. So, of the ₹600,000 required, ₹300,000 needed to be crowdfunded. Against this target, they succeeded in raising ₹314,802. As of August 2016, Milaap, one of the prominent players in the Indian crowdfunding market, had raised over $12.7 million through donations and microloans.[6] Spread across almost 50,000 projects, it has averaged around $260 per project. The beneficiaries of these low-cost loans and relatively small donations have largely been people from rural areas and the underprivileged sections of society. Crowdfunding has also been used for social causes or raising donations for individual problems (an accident victim, for example).

Another new use is peer-to-peer (P2P) lending. Platforms have been developed for this and some of them aggregate incoming funds into a common kitty and use that to lend— almost like a bank. This portfolio approach reduces risks for the lender by effectively spreading the amount across many borrowers. At the same time, these models tap into the surplus cash—even small amounts—that are available with individuals, thereby making for better circulation and utilization of financial assets. The platforms do due diligence on the borrowers, based on data provided by them or culled from public sources.

Such analysis—seeking to rate the creditworthiness of an individual—is done far more extensively and rigorously by companies that have recently come up to extend short-term, small-amount loans to individuals. Typically, these are bridge loans till their next salary. This business would not have come up or been viable but for the ability to mine a great deal of data about an individual, and then analyse it really fast. Much of the data is tapped from social media and is, therefore, unstructured. Sophisticated data analytics and algorithms help to assimilate and analyse this. The result is a decision, taken within minutes, about whether the individual should be given the loan requested.

It is expected that the P2P-lending sector can make a significant contribution towards financial inclusion. To facilitate this, RBI has brought the sector under its regulatory ambit. P2P-lending platforms in India include companies like i2iFunding and Faircent.

Loans to small entrepreneurs or micro-enterprises will be greatly facilitated by emerging trends and new technologies. As digital payments become more common, it will be possible to tap into a database of these, to track the transactions of micro-businesses (a tea stall or vegetable vendor, for example). This, in conjunction with other data, could help determine the credit-worthiness of a business. Loan size, period and interest rate could be tailored accordingly and special products (credit packages) could be created to meet the needs of similar entrepreneurs. To facilitate this, work on a platform called 'IndiaStack' is on. A combination of bank account, mobile and Aadhaar (the unique ID) will link the identity verification with the means of transaction (mobile wallet) and income and expenditure data (through the bank account), which will make presence-less and paperless transactions possible anywhere, anytime. The resulting database can be used for credit-risk assessment. This will open up vast opportunities for micro-businesses to tap into credit for their growth.

Data analytics and algorithms are also being used by banks to decide on consumer loans. Some banks say they can convey a decision on a loan within a few minutes of submission of relevant information by the customer. Data analytics has, for years now, been used by credit card companies to detect possible frauds. Past data on customers enables trends and patterns to be discovered; any sharp deviation raises a red flag, triggering a more detailed look and, if required, a call to cross-check with the customer about the transaction.

However, many of these applications raise concerns about issues related to privacy, especially regarding sensitive personal financial information. For example, in August 2018, Pune-based Cosmos Bank lost ₹94 crore in a coordinated digital fraud comprising thousands of online transactions, carried out through a malware attack on the bank's systems.[7] Such cases emphasize the need for strong data protection and privacy laws. Work on this is already under way in India, and the committee chaired by Justice Srikrishna has given its recommendations to the government. Hopefully, the resulting regulations will provide adequate safeguards, while allowing innovative applications to be implemented.

WHEN TECHNOLOGY OVERTAKES REGULATION

In stock exchanges, high-speed computing combined with sophisticated algorithms enable traders to take advantage of even the smallest lags in stock prices between different exchanges/ trades. This high-frequency trading (HFT) has caused concern and many stock exchanges have taken steps to curb it, because brokers who have their terminals in the stock exchange building are able to access data and execute orders in that split second, which enables them to benefit from the time lag. Also, bigger traders with faster computers have an unfair advantage over others.

This is but one more example of how technology overtakes laws and regulations, necessitating ever newer ones. Sometimes, archaic regulatory frameworks constrain the rollout of beneficial innovations; at other times, new technology spurs knee-jerk reactions leading to new and constricting laws. Globally, the financial sector has been a highly regulated one, and bankers are, by nature, a cautious and conservative breed. While this mindset certainly slows down innovation, one does begin to appreciate the value of careful and calibrated change. Such caution was one reason why India weathered, with minimal impact, the post-Lehman storm[*] that raged worldwide.

The financial crisis was catalysed by sub-prime lending, when high-risk loans were made and then cleverly packaged together into derivatives, which were sold in the market. It is worth noting that much of this packaging was done using sophisticated models that were developed through advanced computer software. This is just one example of how innovations and applications of new technology need not always be beneficial and benign. Therefore, this is one more reason to heed the caution of financial regulators.

At the same time, conservatism should not result in stifling or disincentivizing innovation. Amongst the new approaches towards encouraging innovation and yet being cautious, is the use of regulatory sandboxes. Here, limited relaxation (with respect to time and/or space) of regulations can be tried out

---

[*]This was the widespread financial meltdown, especially in the United States (US) and Europe, which was triggered by the collapse of the large US financial services firm, Lehman Brothers, in 2008.

and tested, before a decision is made about the large-scale rollout of an innovation that requires regulatory changes. Although this concept has not yet been introduced in India, there is discussion about experimenting with sandboxes.

Despite the constraints, the financial technology sector has been amongst the most active and innovative ones in India. The list of new technologies introduced and deployed on a large scale across the country, over the last two decades, has been extensive and impressive. More recent innovations—many based on apps for mobile phones—have contributed to greater financial inclusion, making it easier for the poor to be part of and benefit from the financial system. Digital wallets, mobile ATMs, online banking and a host of other applications have rapidly integrated the large numbers that were left out of the financial system. Direct Benefit Transfer (DBT)—through which subsidies, pensions, scholarships, etc. are credited directly into the bank account of the beneficiary—has made full use of the linkage between Aadhaar and the bank account of the beneficiary. This follows the opening of over 300 million bank accounts as part of the Pradhan Mantri Jan Dhan Yojana (PMJDY)[8], bringing a record number of individuals into the banking system. By seeding each of these accounts with an individual's Aadhaar number (and the mobile number, in many cases), the stage was set for tapping the synergies inherent in the JAM (Jan Dhan, Aadhaar and mobile) trinity. It is interesting to see how the sophisticated technologies inherent in Aadhaar and mobile telephony link into the traditional banking system (now also tech-driven, including, at its heart, the core banking system). Together, they create the JAM trinity, which is already being used for a whole range of purposes, beginning with DBT.

PAPERLESS CURRENCIES

Technology has transformed the financial sector in ways that could not even be imagined when I was growing up. 'Anywhere, anytime cash' was itself a revolution for my generation. Now this has long been overtaken by equally (or more) revolutionary concepts like digital cash, mobile wallets, instantaneous cash transfers (even across borders) and cryptocurrencies. The last is not a 'fiat currency'[*] and is not backed by any central bank or country. It is a global currency, backed by the trust and faith of people who trade or invest in it. Every transaction is recorded in a distributed or decentralized digital ledger, and is unalterable. It runs on a new generation of software technology called the blockchain, which represents an innovative approach.

The biggest change is the very concept of cryptocurrencies, which are paperless and outside the control of any country or central authority. Naturally, this gives rise to concerns, not only amongst central banks, but also within governments. Globally, authorities worry about the anonymity inherent in cryptocurrency transactions, which makes them ideal for illegal use, by money launderers, drug dealers or terrorists, for example. Also, they tend to be highly volatile, driven by speculative surges. Bitcoin, for example, saw its value increase by a whopping

---

[*]Currency that is declared as legal tender by the government, but is not backed by a physical commodity.

1,400 per cent in 2017. Regulators around the world are, therefore, wary of cryptocurrencies, and many have issued warnings about them. In India, both RBI and the government have issued cautionary notifications asking people to be wary of investing or transacting in these currencies. Many other countries, too, have cautioned or warned people about the dangers and uncertainties of investing or using cryptocurrencies. Some are now thinking of issuing their own cryptocurrencies, controlled, like traditional currency, by the central bank. Of course, many see this as ironic: a major USP of cryptocurrencies has been that they are *not* centrally controlled.

Many people around the world are investing in cryptocurrencies to make quick and huge gains—and some even have. Many, however, have incurred big losses. This reminds me of the craze in India, around three decades ago, of investing in tree plantations. Many companies sprouted with the implied promise of large future profits (with no tax, as it was agricultural income and there was high demand for this rapidly decreasing natural resource). Well-intentioned friends—quite knowledgeable about the right investment strategies—advised me to move whatever little savings I had from fixed deposits in banks to shares of these plantation companies. Under pressure to not appear financially naive and obsolete, I did make investments (fortunately only a small part of my small savings) in these companies. Like most others who got taken in and invested, I, too, ended up burning my fingers. Soon, these plantation companies wound up, and one no longer hears of large organized investments in this area.

The year 2017, though, saw a spike in cryptocurrency investment, with their values skyrocketing. Bitcoin, for example, was valued at $915 in January 2017 (the average closing price through the month) and reached an astronomical $16,860 on 6 January 2018. Bitcoin is, of course, the dominant and best-known cryptocurrency, with about 54 per cent market share. However, some of the lesser-known cryptocurrencies did even better: for example, Monero quadrupled to $349 in the last two months of 2017. One reason for this may be that Monero is amongst the currencies most favoured by ransomware attackers. Such attackers typically lock the computers of victims (individuals or organizations)—in effect holding their computer 'hostage'—till they pay a ransom. Now, these criminals demand payment in virtual currencies, with Monero being amongst their favourites. This is because Monero, like some other virtual currencies such as Ethereum, is designed to avoid tracking. This is part of the ongoing battle between criminals and law enforcers, where each tries to stay a step ahead of the other. A number of analytics firms are improving their ability to identify digital cash troves linked to crime or money laundering, and are alerting exchanges so as to prevent the conversion of virtual currencies into traditional cash.

Bitcoin is no longer the favoured cryptocurrency amongst criminals because the underlying blockchain technology records the Internet addresses of the 'send and receive' transactions, as also the amount and time. Clearly, the fact that these can be used as irrefutable evidence worries criminals. A careful watch on these and other transactions can help to keep track of the fund movement and ultimately nail them. Some of the other cryptocurrencies offer various other security features. For example, Zcash ensures privacy protection by encrypting the sender's address. This renders near impossible the task of identifying a sender by looking for correlations in addresses used in multiple transactions so as to pinpoint the real one.

For some reason, South Korea has emerged as a major location in the trading of cryptocurrencies. Such is the extent of trading and the demand for these currencies there that they trade at a premium of about 30 per cent, compared to other countries. Given this, and the growing concern about cryptocurrencies, the South Korean government, on 11 January 2018, announced its plans to ban cryptocurrency trading, following which authorities raided local exchanges for alleged tax evasion.[9] The South Korean Justice Minister, Park Sang-ki, announced that the government was preparing a bill to ban the trading of virtual currencies on domestic exchanges. This led to a plunge of as much as 21 per cent in the local price of Bitcoin. Apparently, this sharp market reaction led to back-tracking by the government, with the President's office saying that the ban had not been finalized.

The saga in South Korea is indicative of both the demand and volatility of cryptocurrencies. Sudden and substantial spurts or drops in their value are clear pointers to the extent of speculation in these currencies. Even a cursory look at the volatility in value justifies the cautionary notes and warnings issued by various central banks.

While cryptocurrencies have clouds hanging over them as far as regulators and governments are concerned, the technology—

blockchain—that serves as their underpinning, is a different matter. With its many advantages, blockchain can be used for a variety of applications beyond the financial sector. The Internet has led to a great deal of disintermediation: cutting out brokers, agents and other intermediaries such as travel agents and real estate brokers, for example. Now, blockchain does away with the need for an intermediary when exchanging value. It secures all information about creation, transactions and validation of cryptocurrencies, in an encrypted or cryptographic form (hence the generic name 'cryptocurrency', for all these forms of money). Interestingly, the real name of the founder of Bitcoin—the most prominent of the cryptocurrencies—is not known, though he had identified himself as Satoshi Nakamoto, and published a paper[10] in 2008, which detailed the concept.

Blockchain is now finding a variety of uses. The banking sector itself uses it for various applications. Beyond this, it is being used for land records, property sales and other areas where its unique characteristics of creating, transacting and validating information are critical. These functions are traditionally played by a trusted intermediary (the government, for example) but they have the drawback of potential corruption or manipulation. Blockchain practically eliminates these possibilities. In India, apart from various industries, the Central government and state governments are actively looking at how best to leverage blockchain for their needs.

It is still too early to say whether cryptocurrencies represent the wave of the future, or are merely one more passing fad that will soon disappear.

NEW OPPORTUNITIES, NEW DANGERS

As one looks at the rapidly changing technological scene, the future seems to be full of new possibilities. Will cryptocurrencies supplant traditional ones? Will AI-based credit ratings, depending on past behaviour and modelling, deprive some people of credit—forever? Will the Internet of Things (IoT) data on driving skill and style, influence behaviours (for example, more careful driving) because the data is used to determine insurance premia? Many such issues are thrown up by the technologies already here or on the horizon. In addition, there will certainly be developments that we cannot foresee today. In short, we are clearly looking at an uncertain future, with the only certainty being that it *will* be different.

The outlook is certainly exciting, but there are dangers too. As more and more financial transactions move into cyberspace, the fear of cybercrime is like a sword hanging over our heads. Theoretically, no encryption can ever be 100 per cent safe. Not only are hackers getting more skilful, they are also getting more organized. Teams of hackers, with capabilities and sophisticated hardware, are now known to be operating in organized gangs, some of which are even transnational. In addition to these 'non-State' players, there are dark stories of State-sponsored or State-supported groups, which could well indulge in an undeclared cyberwar. After all, in the modern world, disrupting a country's financial system can be more damaging than bombing a power plant.

In such cyberattacks—whether by individuals or organized groups—the financial sector is clearly a prime target; hence, the need for special measures to ensure its invulnerability. In this too, most of the 'protective armour' has to come from technology.

Despite these concerns and dangers, technology is clearly set to play an ever increasing—even dominant—role in the sector of money and finance. With online and digital transactions growing rapidly, and connectivity being critical, banks have had to invest heavily in acquiring IT capabilities. Sensing an opportunity here, some telecom service providers and mobile wallet companies have moved into banking, given their inherent capabilities. In India, for example, Airtel and Paytm sought—and have been granted—banking licences (presently as 'payment banks'). At the same time, Amazon and Google have also entered the payments space with Amazon Pay and Google Pay, respectively.

The broad trend is quite clear: tech companies are entering the financial sector, and in a big way. New business models, too, are emerging, with new players bringing in a different perspective and innovative ideas. Will this mean the end of traditional banks? In particular, is the brick-and-mortar bank nearing its death? It seems certain that they will have to reinvent themselves to survive. What this new avatar will be is still unclear. The only certainty is that profound change is in the air.

2

DR GOOGLE'S CLINIC

How the Internet and Technology are
Transforming Healthcare

Over centuries, in a slow and subtle manner, technology has caused a big change in one of the most important aspects of human life—the duration of life itself. In India, the change, over the last few decades, has been more dramatic: life expectancy at birth was 44.36 years in 1965; half a century later, in 2015, it had increased by over 50 per cent, to 68.3 years, with Kerala topping other states, at 75.2 years[11]. One result of the increased longevity is that there are so many more elderly people (defined here as those above sixty years). According to a 2017 United Nations (UN) report, their proportion in the country's population is expected to further increase from 8 per cent in 2015, to 19 per cent in 2050, with the absolute number tripling from 100 million to 317 million[12]. The Indian government now estimates the figure at 340 million[13].

This has had a profound impact on society and the healthcare system, the implications of which are discussed later.

Most people are living longer for a variety of reasons, including adequate food and better nutrition. However, the primary reason is clearly the improvement in healthcare. Preventive technologies like vaccinations, and curative ones like antibiotics, have been the key elements, along with diagnostic tools like blood tests and X-rays. Over the last century, these have seen steady progress, thanks to research and technological development. At the same time, a range of new technologies have been developed for diagnostics as well as treatment. Some of these are simple, but can have great impact. One example is oral rehydration therapy (ORT), an easy, self-administered way of handling diarrhoea. The wide prevalence of diarrhoea in India has been one reason for poor growth amongst children and a cause for high infant mortality. While ready-made ORT packets are available, this is a remedy that can be made at home too; in fact, it has been traditionally known in one form or another. At the other extreme of sophistication are high-tech tools like magnetic resonance imaging (MRI) and computer tomography (CT) scans used for diagnosis, or radiation equipment used to treat cancer. These are expensive and complex equipment, with highly trained people required for operating and maintaining them.

India's progress with regard to some health indicators is quite impressive in terms of actual numbers, especially when compared to the pre-Independence era. However, in terms of global benchmarks, and as compared to most other countries, it still comes out poorly.

PUSHING THE FRONTIERS

The convergence of electronics, computers and software, with the broad discipline of medicine, has resulted in a host of new devices that are radically altering healthcare. First-generation electrocardiography (ECG) and ultrasound equipment is being supplemented—and sometimes even replaced—by cheaper and better products, including portable and low-cost ECG and ultrasound devices. The reduction in cost and the portability of these devices have resulted in extending their use to a wider section of the population and to those living in rural areas.

Digital X-rays and software-driven image processing techniques are improving accuracy and enabling better diagnosis. New technologies like MRI and CT scans epitomize the combination of electronic hardware and computer software with medical expertise. These are large, sophisticated and expensive pieces of equipment, but they have made diagnosis so much more incisive, resulting in a big impact on healthcare. These facilities are no longer restricted to large hospitals, and there is already a rapidly growing number of stand-alone 'imaging centres' (especially in larger cities). It is doubtless that advancing technology will reduce the cost of equipment and make these facilities available on an even wider basis.

Apart from these facilities, which require specially trained or skilled people for their operation and especially for their maintenance, new technology is helping to create easily usable equipment. Devices like digital thermometers and blood pressure measurement instruments have been around for a while. Now, digital blood sugar measurement devices, requiring just one drop of blood for analysis, are also common. This makes it possible for patients to monitor their sugar levels themselves, at any desired frequency, without the need to visit a laboratory or hospital. New devices that are simple and cost far less than conventional methods have been developed for ophthalmological applications also (for example, checking for glaucoma or cataract). Besides the above, portable and lab-in-a-box equipment now enable testing and diagnosis for a number of key ailments at a low cost. This could potentially revolutionize healthcare for those in remote and rural areas.

A question that is often asked is: 'What after diagnosis?' Some of the devices are simple enough to be used by a patient or an untrained caregiver, but what follows? The user can access the sugar or haemoglobin level on his/her own, but who will analyse the result and suggest the action (or medication) to be taken? These questions are particularly pertinent in rural areas, where the physical infrastructure for healthcare is poor and even those facilities that are there, are hugely understaffed. Most rural health facilities (right from the sub-centre and primary health centre [PHC]) do not even have the minimum health staff that they are sanctioned. Given the shortage of qualified medical personnel—not only specialists, but even general practitioners and nurses—this is unlikely to change soon. Compounding this is the fact that doctors and nurses are reluctant to work in rural areas, and those positioned there are frequently absent. In this context, is mere diagnosis enough?

A NEW HORIZON FOR PUBLIC HEALTH

In the above circumstances, telemedicine offers an alternative possibility, at least as an interim solution. It may not be as ideal as an on-the-spot specialist or doctor, but till such time as that ideal is realized, telemedicine can play an important role. It is, of course, dependent upon ICT and appropriate—and more importantly, reliable—connectivity between rural areas and specialized hospitals. A number of experiments and pilot projects have been carried out, with remote/rural areas linked to specialists in one or more hospitals. In most of these cases, the standard practice is similar. The first step is to record the patient's personal data, including sex, age, height, weight, past medical history, etc. These can later be pulled out, with the patient's permission, from an electronic database where they are stored. Next, a paramedic at the remote location helps in getting the patients' data from the diagnostic device, which is then transmitted—either automatically or with the intervention of a paramedic—to the hospital (which could be hundreds or thousands of kilometres away) through a communication link. There, a specialist analyses the data and diagnoses the problem, and may then ask the paramedic or the patient for additional data. Following this interaction, the specialist will advise on the action—referral to a hospital, medication, re-check after a given period, or nothing at all.

I recall, from my days in the Indian Space Research Organisation (ISRO), what was probably the first telemedicine experiment done in the early '80s. This used the Indian communication payload on-board the APPLE satellite launched in 1981 (the odd name is an acronym for Ariane Passenger PayLoad Experiment; Ariane being the launch vehicle). It was used to link doctors in a hospital in Ahmedabad with specialists at the All India Institute of Medical Sciences (AIIMS) in Delhi, enabling them to discuss cases and seek guidance from experts.

Since those days of transmitting a few rudimentary parameters, telemedicine has made great progress. First, at the remote or patient's end, the kind and amount of data that can be collected has grown vastly, thanks to new diagnostic equipment; second, the quality and bandwidth of the communication channel is a lot better, enabling the transmission of images and videos; third, small portable terminals make it possible to link remote places via satellite (ISRO recently deployed a telemedicine facility in Siachen); finally, at the hospital, AI and high-powered computing, backed by massive amounts of data, enables doctors to diagnose the ailment with greater speed, accuracy and certainty. This makes it possible to fully operationalize such a system. However, a full-fledged and large-scale telemedicine network is yet to take shape in India.

With rapidly improving broadband connectivity and lower costs, there is little doubt that the proliferation of ever better self-diagnostic devices will lead to extensive use of telemedicine. Possibly, the government will upgrade each PHC and sub-centre into telemedicine access points. It is also likely that private telemedicine networks may be set up by large hospitals or by independent service providers. Given the spread of urban centres, it is not inconceivable that we may see such centres as 'neighbourhood facilities' in big cities. Remote consultations may also replace the (now rare) home visit by the doctor.

UNDERSTANDING TECHNOLOGY ENABLEMENT

One result of new developments is that doctors have to now understand a great deal more about technology. In fact, progress in this area depends very much on such understanding, for it is only then that they can make demands on and challenge technology to do those things that are high on the list of healthcare priorities. This calls for a revamp of medical education, which must now necessarily include a fair amount of exposure to technology. Equally, the training of paramedics will have to change, so as to develop their capability to operate various sophisticated high-tech devices. The invasion by electronics means that medical facilities now 'smell' more of silicon than of formaldehyde, and resemble a high-tech IT centre rather than a traditional hospital.

If hospitals have changed, so has the doctor's consulting room. For people of my generation, the image of the consulting room of the past is embedded in their memory—the doctor behind a table, with a stethoscope around his neck; the mercury-filled blood pressure monitor on a side table; a syringe somewhere in the picture; a steel stool for the patient, beside the doctor; and a bed for examining the patient further, if necessary. Sadly, as time went by, some doctors had no time to listen to the patient's complaints, examine him/her and discuss the diagnosis and possible line of treatment. Metaphorically, the stethoscope remained, but the pen became the main tool of their trade. With the rush of patients, all the doctor had time for was a cursory hearing of the patient's woes; at best, a very quick examination; and then the wielding of the pen to prescribe medicines or (increasingly) various tests.

In more recent times, an addition to the doctor's side table has been a computer, and the mobile phone has practically displaced the syringe. In urban India, a doctor's consulting room has moved from his/her residence to a larger hospital. Of course, the personalized attention of a doctor who knew the patient and his/her medical history is now but a faint memory, and as mentioned before, is as rare as a home visit.

If the doctors, their approach and setting have all substantially changed, so have the patients. 'Technology enablement' is not limited to the doctor; the patient, too, has learned to leverage technology. Thanks to a previous job, which has indelibly linked me with the IT industry, I am at the receiving end of many a doctor's ire. They hold *me* responsible for the fact that many of their patients, being computer-savvy, have read up everything possible about their ailment on the Internet. They think that their knowledge exceeds that of the doctor, and they cross-examine the latter on every aspect of their ailment and the prescribed medication. They know all about the side-effects and after-effects of the medicines and whether any have been banned anywhere in the world. They also read up on the very latest research findings in this field and like to show that they are, in fact, one step ahead of the doctor. As one doctor told me, he's very tempted to tell them to go to 'Dr Net' or 'Dr Google' for consultations, instead of coming to him.

As a matter of fact, some are doing precisely that: for what they perceive as 'non-serious' ailments, they self-diagnose using information from the Internet (which also suggests the medication). As long as the suggested medicines are available over the counter (meaning that the chemist is willing to sell them without a doctor's prescription), they self-medicate. Whether such use of technology is appropriate is both a matter of debate and concern, but it has begun to happen. It is doubtless that as access to doctors gets more difficult and expensive, and as medical sites on the Internet get better, this will become a more common trend. This will also be accelerated by entrepreneurs— many of whom may possibly also be doctors—who see a potential business opportunity and, so, invest in developing more sophisticated sites based on expertise, large amounts of data drawn either from patients or crowdsourced, and data analytics.

Does this portend the end of the road for general practitioners (GPs) at some point in the near future? Is this one more job that technology will make redundant? Certainly, a considerable reduction in the demand for GPs is likely in urban areas. Rural India is so underserved that demand there will continue to grow for a while, though one should not rule out the possibility of villagers leap-frogging from the present state directly to the Internet stage, especially if the websites are in a local language and are user-friendly. However, there is another possibility: in this age of machines, people sometimes crave a human touch (think of your experience with interactive voice response systems, where you sometimes desperately want to talk to a human being rather than a disembodied voice). Doctors could provide this, if they reinvent their role. A reversion to older, more leisurely times, with that personal touch of a 'family doctor', may well rejuvenate the institution of the GP. If not, it may head towards extinction.

Technology is also being used as a platform for reach, connectivity and better use of scarce resources. One example is platforms that connect patients to available doctors, facilitating both appointments and remote consulting. This 'uberization'* of medical services makes fuller use of doctors who may have spare time, while allowing them to set their own schedules. Patients, too, can plan and optimize their schedule by seeking an appointment at a time that best suits them. This makes it mutually convenient for both parties and obviates the need for travel. Payment of the doctor's fee is made online, and additional functionalities (like uploading the patient's medical data for access by the doctor) increase efficacy. In a sense, this combines the advantages of seeking 'advice' from a computer—any time, from anywhere and without travelling—with the more traditional approach (consulting a doctor), thus leveraging technology but not losing the human interface.

WHEN SCIENCE EXCITES, BUT ALSO SCARES

What has been discussed so far is mainly the 'external' use of technology. However, for many decades, technology has been used 'internally' too—*within* the human body. Add-ons attached to the body have been common for a while—hearing aids, prosthetics/artificial limbs, dentures and spectacles. The last two

---

*A neo-euphemism derived from the transport services company, Uber.

have evolved to be completely internal—spectacles into contact lenses, and dentures into tooth implants. Metal supports (rods and screws) that hold broken bones together are permanently inside the body, and quite common now. More complex devices, like pacemakers for the heart and cochlear implants for those hard of hearing, have also been around for many years. The latter convert audible signals (or sound) into electrical signals, which are then relayed to the brain. While this is revolutionary, the number of people with cochlear implants is still only a few hundred thousand; clearly, cost and other factors are barriers.

The idea of so-called 'bionic beings' or 'cyborgs' has been around in science fiction for quite some time. Amongst these, the one closest to reality—and one of the best known—is probably Luke Skywalker's bionic hand in the superhit *Star Wars* movies. A true bionic hand, to mimic the human hand, must not only be fully controllable but should also have sensors that can differentiate between soft and hard objects. A big challenge to designers had long been sensors that would enable a bionic hand to pick up a delicate glass without crushing it. Such a hand has now been developed.[14] Unlike the first such device developed in 2014, which required large equipment, the 2018 version is portable[15]. Messages from sensors in the bionic hand are linked to a computer (which can be carried in a backpack), processed and sent to the brain through tiny electrodes implanted in nerves in the upper arm. This has been tried and works well, with the patient able to correctly differentiate between hard and soft objects, even while blindfolded.

This trend, of a machine replacing a part of the body, is surely going to be taken much further, as both technological capability and our understanding of the human body progress. Machines that carry out key body functions already exist—for example, a heart-and-lung machine or dialysis equipment—but are too large or insufficiently evolved to fit into the body. However, it may only be a matter of time before they do.

Besides the above, more exciting developments are taking place in other fields. Work in the area of stem cells is making it possible to 'grow' various organs. Already, there are successful instances of growing human organs with animals as surrogates.[16] Some countries have regulatory constraints on stem cell research, but others are pushing ahead. Certainly, there are ethical questions and dilemmas, but many feel that the prospective good outweighs the potential dangers.

Another matter of ethical concern and dilemma is developments derived from genomics. In particular, techniques of gene editing have progressed to the point of becoming operational.[17] The ability to remove or add genes in the basic DNA structure makes it possible to manipulate this building block. While the positive aspect is that this could enable the prevention of hereditary diseases, the technology could also be used for other purposes. Some are already speaking of creating 'superhumans' by manipulating genes to enhance certain features, or producing tailor-made babies. The possibility exists that in the future, someone may 'order' a baby with a certain skin or eye colour, a more defined nose, and so on. One is reminded of Adolf Hitler and his obsession with the 'pure' Aryan race. Eugenics is not new to biology, and Hitler gave it a sinister and racist touch, but now it is nearly in the realm of reality. This is certainly one area in which science excites, but also scares.

Gene editing and splicing can do more: they could, for example, create completely new life forms—species that have never existed in the past. It is these possibilities that have given pause to many, especially in the scientific community, and a lot of voices are being raised, asking for self-restraint by scientists regarding their 'innovations'. Some are even arguing for more formal control by way of a regulatory regime by governments, even extending to an international agreement.

New technologies always pose new challenges, even as they open up vast opportunities. Genetics is no different, except that it is highly sensitive, given its immediate and significant impact on human beings. The issues concerning regulation—whether to regulate, what to regulate, to what extent to regulate, who is to regulate, and when to regulate—are not uncommon in any area of technological innovation. However, in this case, we have the added dimension of ethics.

NO LONGER A PIE IN THE SKY

Like genetics, neurotechnology is another new frontier. Despite much research, our understanding of the brain is still limited. Even so, ample progress has been made in the last few years, and this—in combination with the rapid growth in computer technology—has led to a number of new devices based on brain–computer interfaces (BCI). These are offering new hope to people with serious disabilities. A recent example is of a patient paralysed below the shoulders due to an accident.[18] He can now use his hand to feed himself, thanks to electrodes implanted in his arm, which stimulate the muscles. But the wonders of new technology go beyond that: he can now control his arm through the power of thought! The intention of moving his hand is reflected in neural activity within his brain's motor cortex. Implants in his brain detect the signal and process it into commands to activate the electrodes in his arm. This decoding of thought was pure science fiction even till a few years ago. Today, researchers are able to tell what words and images people have heard and seen, from neural activity alone. Surely, this is magic!

The possibilities being opened up by BCI are becoming evermore extensive as the pace of research in this area quickens. Facebook is talking of thought-to-text typing; Elon Musk, along with eight others, has set up a company called Neuralink to develop BCI; and the US armed forces are also showing great interest in this area. These are indicative of its immense potential, as well as a guarantee of rapid progress. Certainly, the augmentation of human capabilities—a strong potential use of BCI—is of considerable and immediate interest to the military. As is well known, technologies of interest to the military are developed faster, thanks to generous funding and the imperative to be a step ahead of rivals. Therefore, one can expect rapid progress in this area.

An application that may fructify sooner than others—in part, because of its commercial potential for games and robotic manufacturing—is the 'skull cap'. The constraints of non-invasive electroencephalogram (EEG), which is its difficulty in picking up high-resolution brain signals through layers of skin, bone and membrane, are being overcome. EEG caps can now be used, for example, to create virtual reality games, using thought alone. This can also be used to control industrial robots. Nissan Motor is testing its 'brain-to-vehicle' (B2V) technology, unveiled in the International Consumer Electronics Show in Las Vegas in January 2018.[19] The B2V system uses a skull cap to pick up and transmit brain signals to the steering, accelerator and braking systems, which start responding before the driver initiates action. While the driver is still steering, the car anticipates these movements and begins to take action 0.2 to 0.5 seconds sooner. Nissan's researchers insist that the skull cap is not reading the driver's mind but only decoding brain activity just before a voluntary action. To a layman, though, it does seem very much like mind reading.

However, operationalizing applications of BCI is not easy. It is one thing to do it in a research laboratory and quite another to convert it into a viable product. Size, complexity and costs are all barriers. Our understanding about the functioning of the brain is yet limited. We will need many more advances in both neuroscience and technology to overcome the obstacles before BCI results in commercially viable or extensively used devices. Even so, around 150,000 people already have electrodes implanted in them to provide deep brain stimulation as a means to help them control Parkinson's disease.[20] While it may take some time, there is little doubt that in the foreseeable future (probably sooner, rather than later), brain implants and BCI will be extensively used.

As new developments take place in this field, there are also growing ethical concerns. In a sense, the ability to decode thought has been established. Scientists have also demonstrated the ability to 'inject' data into monkeys' brains, instructing them—via electrical pulses—to perform actions. This is certainly scary. It opens the possibility of implants being hacked and the individual being 'taken over' by the hacker. Even the positive aspects—for example, enhancing cognitive abilities through implants—have possible downsides: given the high costs involved, this may become a source of further inequality and elitism. Instead of the smart becoming rich, the rich will become smart—thanks to implants that enhance their 'brain power'. As we move up the ladder of technological complexity with regard to BCI, ethical questions assume greater importance and relevance.

To most in India, much of this may seem meaningless. Brain implants and other high-tech health innovations appear to be a pie in the sky to those for whom even the simplest form of healthcare is not available; it's like asking them to eat cake when even bread is unaffordable. After all, India's key health indicators are as bleak as those of the most backward and poor countries, and are generally amongst the lowest in South Asia. Despite the improvement in the under-five child mortality rate, from 43 per 1,000 in 2015, to 39 in 2016[21], India recorded the largest number of deaths of children under five years of age (802,000)[22]—a world #1 rank that no one would want. Its record on infant, child and maternal mortality is dismal, not only in absolute terms and percentages but also in comparison with countries having similar per capita incomes. Within India itself, there is a very large spread between states that are doing reasonably well in healthcare versus those who are not. For example, Kerala has some health indicators that are at a level comparable to advanced countries. Much of this has little to do with technology in a direct sense. However, it is certainly likely that the use of technology will further enhance the health status. Increasing awareness (for example, through mobile phones and social media); data collection and monitoring; improving cold chains for vaccines and medicines; maintaining and updating health records; automated reminder systems for taking medicines, visiting a health centre or getting vaccinations—technology can contribute to the improvement of all these basic elements, leading to an improvement in health. This is quite apart from the possibilities of using technology for cheap and quick diagnosis, with minimally qualified paramedical staff, and of new technologies at the curative and care stages.

New technologies and the growing range of their applications promise to completely transform not only healthcare but human life itself. This is certainly amongst those areas where technological disruption will be maximal.

3

TO SILICON SIR, WITH LOVE

The Machine as Teacher

Many consider that the world is evolving towards a 'knowledge economy'. However, this is not strictly accurate. In ancient times, it was the knowledge of how to hunt better or bigger animals, and how to find the location with the best fruit, that propelled the hunter-gatherer economy. Later, it was knowledge of agricultural practices and how to increase productivity that helped the agrarian economy. The era of global trade began with discovering overland routes (e.g., the Silk Road, which made possible trade from China in the east, to as far as Europe in the west). Later, the monsoon ('trade') winds, navigation and ship-building—all dependent on knowledge—provided an impetus to economies. The Industrial Age was more obviously linked to knowledge, as invention and discovery quickly translated into equipment and machinery.

From this perspective, economies have always drawn from,

and thrived within, knowledge. A 'knowledge economy' may, therefore, be a misnomer for characterizing or differentiating our times, especially if the phrase is interpreted as meaning an economy that is dependent on knowledge. What is different, though, are two important factors: the speed at which 'new' knowledge is being created, and the proportion of economic value attributable to new knowledge (of, say, the last five years). The latter varies vastly, from sector to sector, or service to service, but, in general, the contribution of such knowledge is far greater than it has ever been in the past.

The speed of knowledge creation is certainly far more today, and is driven by various factors. One is competitive pressure between companies and countries, as also between researchers. Another is better communication, which facilitates long-distance collaboration as well as near instantaneous information exchange. The third is a shortage of resources—for example, depleting natural resources are giving rise to research on new materials or alternative technologies, to reduce their consumption. Knowledge creation is also driven by other emerging concerns: for example, concerns about pollution and climate change are driving intensive research on alternatives to coal and certain refrigerant gases.

The knowledge value embedded in products or services today, and the pace of knowledge creation, are unparalleled in history. To that extent, if we define these as the characteristics of a knowledge economy, then we have evolved to that stage. Added to this are two other factors: the speed with which knowledge is applied (that is, its conversion into products or services), and the rapidity with which new products penetrate day-to-day living.

TECHNOLOGY AND ITS IMMENSE POSSIBILITIES

Clearly, then, irrespective of semantics and labels, knowledge is a vital element of the economy. Today, in conjunction with its derivative (technology), it is widely recognized as the fourth factor of production—after labour, land and capital. In fact, in many cases, it is not only more important than traditional factors but can sometimes replace one or the other of them.

Given this pivotal role, every organization and government is seeking ways of growing its knowledge base as quickly as possible. Since knowledge is derived from research, and the most important input for research is education, it is necessary to build a big, strong and high-quality base for education. Even where knowledge or research outputs are acquired from elsewhere, absorbing and using these require a certain level of education. This, then, is the overwhelming economic argument for the importance of education, apart from it being important in itself.

It is in this context of the links between economic growth, knowledge, technology and education that one needs to look at where education stands today, and its possible course in the future. There is, of course, the other (and as some may say, more important) dimension. This is the place of education in the broader framework of social and cultural evolution. A modern society is defined as much by its civilizational values as by its economic well-being. Many would consider the two as being interlinked, though, unfortunately, a few countries still defy such a correlation. Yet, there is little doubt that literacy and high levels of educational attainment are hallmarks of a society that is generally cultured, peaceful and democratic.

Extending educational opportunity to all, ensuring high-quality learning and promoting cutting-edge research are now the ambition of every country. Concrete evidence of the benefits accrued thence are found around the world, with universities—particularly in developed countries—being the powerhouse of new technologies. In many cases, these are also the birthplaces of new ventures, incubating ideas and taking them to commercial success.

In this process, technology is not only the output of universities, it is often also the input. Increasingly, the education system is leveraging a host of technologies to extend its reach, improve its quality, increase its efficiency, and accelerate and personalize learning. For countries like India, with a huge backlog in all stages of education and ambitions to be at the forefront of knowledge, the need for and the possibilities of technology are immense.

Even a cursory look at the figures indicates the magnitude of the challenge facing India: 260 million children in schools (2015–16) and 34 million in higher education[23]. At the school level, while the enrolment in primary classes (age group six to fourteen years) is now a satisfactory 97 per cent, the quality of education is abysmal. A 2016 research report looked at the abilities of the age group of fourteen to eighteen years. It showed that about 25 per cent children in Class VIII could not read Class II texts and that only 54 per cent could answer at least three of the four questions based on the written instructions on a packet of Oral Rehydration Solution (ORS). Further, 57 per cent of them could not correctly solve a simple three-digit-by-one-digit division problem.[24, 25] Therefore, any satisfaction that one might feel at the extent of enrolment is clearly misplaced.

The Sarva Shiksha Abhiyan (primary education scheme) placed a lot of emphasis on getting children into school and particularly on inputs. Classrooms, toilets, drinking water, playgrounds and a specified number of teachers were made mandatory under the Right to Education (RTE) Act. Funds were provided to ensure these and regular monitoring pushed implementation. 'Softer' aspects, like school management committees, were far more difficult to put into place and did not get the same level of attention. Importantly, there was little focus on quality and learning outcomes. The overall result (at least up to now) is clear: while enrolment and physical facilities have improved, learning has not.

The serious problem of school dropouts has not been solved either. The RTE stipulation of 'no detention' up to Class VIII does not seem to have helped, belying the hypothesis that a lot of children dropped out when they failed or were detained in the same class. The well-intentioned and progressive continuous and comprehensive evaluation (CCE) method was poorly—or, in some cases, not at all—implemented, mainly due to the inadequate training of teachers. This, and the lack of proper corrective and supplemental education, led to dropping standards, as students without sufficient learning were automatically moved up to higher classes. As a result, there was a strong and growing group (mostly consisting of parents and teachers) that demanded scrapping of the 'no detention' policy. Finally, in 2017, the Central government decided to do away with the mandatory policy and left it to individual states to decide on this. Time will tell about the impact of this decision, particularly on learning and dropout rates.

BEYOND THE TEXTBOOK, OUTSIDE THE CLASSROOM

For many decades, efforts have been made to improve learning outcomes by supplementing traditional classroom teaching with technology. Educational broadcasting—first, radio, and then television—has long been one such initiative. In fact, TV in India (started in 1959 with an experimental telecast) began with broadcasts for schools and school teachers, when regular daily transmission was operationalized in 1965. The first—and for long (till 1972), the only—TV station in India, located in Delhi, regularly broadcast programmes for schools.

The use of technology for education was taken to a higher level (literally) with the advent of satellites. From 1975–6, the historic Satellite Instructional Television Experiment (SITE) broadcast educational programmes for primary school children in remote rural villages in six states of India, through 'direct broadcasting' via satellite (in a prelude to today's direct-to-home [DTH]). These educational TV (ETV) programmes were broadcast for ninety minutes every day. During the school vacation, a massive teacher-training programme was organized for two weeks for as many as 250,000 teachers. This exemplified the use of the very latest technology (satellite broadcasting) for education. Independent studies pointed to a number of positive results due to the broadcasts: school attendance had increased, and students' interest and attention were high.

ETV broadcasts to schools have continued since then. However, these have faced problems, including the maintenance of TV sets, availability of electricity, synchronization of the topic of broadcast with the specific lesson being taught in each classroom and the teacher's attitude to TV lessons. The style of presentation and the overall quality of the programmes also leave much to be desired, especially when compared to the slickness of commercial programmes that most children watch. As a consequence, the opinion about the success of ETV broadcasts for schools is, at best, mixed.

The advent of the personal computer (PC) added another technological possibility. Anticipating the 'age of the computer', the government initiated a programme called Computer Literacy and Studies in School (CLASS) in 1985. This had the laudable aim of seeking to create a new generation that would be computer-literate and, thus, build a base for developing individual and national capability in computer science and related areas. Though not seen as a great success, some consider CLASS as the genesis of India's subsequent growth and success in the IT industry.

The use of ETV in higher education was initiated much later: two decades after the first school broadcasts. In the '70s, plans were being made for the expansion of conventional TV stations in the country and operational satellite broadcasting was on the horizon. At that time, with almost prescient vision, a small group of people—led by scientist Yash Pal (then director of ISRO's

Space Applications Centre and, later, Chairman of the University Grants Commission [UGC])—feared that television would end up being monopolized by inane entertainment and commercial interests. Therefore, even as the Indian National Satellite System (INSAT)* was being developed, they ensured that the planning included reserved time on the satellite for educational broadcasts, including a slot for university-level programmes. Thus was born the UGC Countrywide Classroom, launched in 1984—even before the universities were fully ready to produce the content. Programmes were also initiated for structured courses offered by the then newly created (1985) Indira Gandhi National Open University (IGNOU). These programmes were aimed specifically at registered students of the university, as they were course- and topic-specific. The Countrywide Classroom broadcasts, on the other hand, aimed at broadening perspectives, with the philosophy of taking viewers 'beyond the textbook, outside the classroom'. Both had a large and dedicated viewership, but obviously could not compete with televised movies or entertainment programmes. So, while they have been successful in their own right despite limited resources—resulting in inadequate 'promotion', and programmes with little slickness or technological polish—the yardstick of 'ratings' (viewership numbers) made them look like failures. As a result, they were given low priority, with arbitrary shifting of broadcast time and no 'marketing', further reducing the viewership. Still, they continue to survive! In fact, with digitization and a big drop in

---

*A satellite system for TV broadcasting, telecommunications and meteorological imaging.

cost of satellite broadcasting, there are now multiple dedicated TV channels for higher education.

Over time, these broadcasts have sought to increase their effectiveness through the use of newer technologies. Thus, interactivity was introduced by adding a 'talk-back' capability, initially through phone lines and satellite terminals. Also, programmes were digitized and stored on servers, facilitating anytime-access. New programmes added more elements of computer-generated graphics, animation and effects. Thus, the effort has been to leverage technology to the maximum extent possible.

DISTANCE LEARNING REDEFINED

Computer technology and widespread Internet connectivity opened up new vistas for education, especially with the availability of broadband connections and the decreasing cost of communication. These made it possible to increase the efficiency of existing educational technologies and created new possibilities, including open courses, remote tutoring, collaborative research and immersive learning.

Globally, the advent of the Massive Online Open Courses (MOOCs) was not just a semantic change from the earlier distance learning, but took it to a new level through the use of contemporary technology. Today, MOOCs represent high-quality courses, often created by top experts, in a wide variety of fields, and are available (generally free of cost) anywhere in the world to anyone with Internet connectivity. Importantly, many of the creators get feedback (based generally on student performance in embedded tests) and modify the content or pace of the course accordingly. With an increasing number of students, such crowdsourcing of feedback, coupled with the latest techniques of data analytics, enables a constant fine-tuning of course material, so as to enhance learning. Many institutions around the world have now integrated MOOCs into their curricula, enabling students to learn at a time, place and pace suitable to them. All this has been enabled by a combination of communication, computer and pedagogic technologies, enhanced by data analytics.

MOOCs, as noted, have been integrated into the learning process in many universities. However, their significance is, at least, as great for those who have completed their formal education. As new technologies are developed and new findings make obsolete what one learnt a decade (or less) ago, there is need for upgradation of one's knowledge. In today's world of rapidly advancing knowledge, and fast-changing consumer psychographics and business models, such updating of one's skills and capabilities is a must. This dire imperative is already facing organizations in many sectors: how to re-train their existing human resource base. The option of layoffs and the recruitment of younger people with contemporary knowledge is not only a social challenge but also impractical for many other reasons. Equally problematic is the solution of sending a large proportion of the employees back to a university or training institution for full-time courses over extended periods. It is here that MOOCs offer an excellent solution—learning while on the job. This 'learning while earning' model is ideal for both the employer and the employee. Increasingly, for a variety of pedagogic reasons, pure online learning is being supplemented by limited face-to-face interaction, in a so-called 'blended learning' approach. The advantage is that this can be done where feasible; where it is not, the online-only approach is reasonably effective.

Blended learning—a 'brick and click' approach—marries face-to-face classroom learning with the online platform. Many universities are now using such a model, making it possible for students to take courses not offered in their institution. Others use it for parts of a course, or sometimes for a 'foundation' course in conjunction with classroom tutorials. In the context of updating the skills of working professionals, such blended learning is now being extensively used in executive education programmes.

Another technological innovation is the gamification of learning. This seeks to use the popularity and excitement of computer games to promote learning. Given the cost of hardware and of creating sophisticated gaming software, this is an expensive proposition. It is not yet clear whether the depth and speed of learning and the economies of scale compensate for the cost, and make such an approach effective and cost-efficient.

Emerging technologies related to virtual and augmented reality (AR) are being used for a variety of different applications. They will, doubtless, contribute to immersive learning and enhance the quality of education. These technologies are being used to a considerable extent for training. One important use is for simulation, with obvious applications including the training of pilots and drivers. In these and other skilling situations, they can contribute in a major way to bringing nearly real-life environments to the trainee, greatly enhancing the efficacy of training. Given the present cost and availability, the use of AR for training in India is yet limited; however, the scope is clearly immense.

NEW DIMENSIONS

Access to data and information is now fast, easy and quite inexpensive. With Internet connectivity, search engines and websites, one can now get information on almost anything, nearly instantaneously. The old days of scouring books and journals in libraries to dig out a relevant piece of information are long gone. Now, all one does is enter the query in Google's search bar and all the information, along with references and related links, is downloaded in moments. Most reports, publications and millions of books are available online, and many can be downloaded for a small fee, if not freely. The significance of this for education, and especially for research, is obvious. Even students in primary classes browse the Internet to find relevant information (though not everyone is convinced that this quick and easy shortcut is good for children).

An important dimension of ICT is its facilitation of collaboration over a distance. Thus, researchers living far apart— even on different continents—can not only share their findings but also collaborate on a common research problem. This is no longer uncommon, and multi-country teams often work together to improve the quality of research by bringing together experts and speeding up the process (by having more researchers and by using the time difference across time zones to 'extend' the working day). Apart from industrial research, such collaborations are extremely useful for research scholars in universities.

Another dimension of the use of ICT in research is for data analysis. Computers have been used for this purpose for some time now, especially when there are large data sets. This has evolved to data-driven discoveries, where new ones may emanate from the study of very large quantities of data. At a more basic level, technologies like data analytics are being used to sift through large-scale data to identify, for example, the major causes (subject, topic or sub-topic) for poor performance of students, so that teaching can be appropriately modified. This is important feedback for course-content designers (of books and learning material), to teachers, and about teachers (to evaluate their performance).

Technology has, for many years, taken the time and effort out of computations of all kinds. The days of slide-rules and logarithmic tables are long gone; in fact, the last two generations probably have no idea of what these are! Now, any type of complex calculation can be done in moments, by pressing a few keys on a computer (sometimes, even on a mobile handset). Not only that: the results—or any set of data, for that matter—can be immediately converted into graphs and charts. This has eliminated the drudgery of spending hours on calculations and freed up students' time for deeper learning or discussion. There is, though, a contrary view: such computations—and the painstaking efforts involved—were essential parts of learning, with the calculations serving to sharpen one's brain power; the use of technology reduces man's capability to do such calculations on his/her own. It may be a while before research can give us a comprehensive answer on which approach is better.

While ICT and especially the Internet, holds great promise, access to computers is a serious constraint, particularly in a poor country like India. Apart from the cost for individual ownership, many schools also cannot afford computers, particularly at the primary level. Further, in many locations, availability of electricity and Internet connectivity are additional problems.

CLASSROOMS OF TOMORROW

So, what does the future hold? As for all things connected with technology, past trends do not help in predicting the future. New devices, technologies and innovations result in discontinuities, ruling out the utility of trend projections. One scenario may be an extension of what we have witnessed in the last few decades: the increasing use of technology, with incremental impact on overall learning, teaching and the broader educational system. Thus, there will be many MOOCs and other opportunities for 'anywhere, anytime' learning, as also the lifelong process of the upgradation of skills or learning new ones. Apart from traditional universities that have adopted these new technologies, there will be many private educational companies in areas such as content, delivery, certification and system administration, who will also make use of them. New ventures may sprout, which do a 'life-cycle management' for individuals, guiding them through all stages of education throughout their lives. Just as medicines and healthcare have evolved from a few knowledgeable people providing home-made remedies to the massive global pharmaceutical and healthcare industry of today, education may move from the 'guru-shishya' model of yore, to worldwide corporate entities. Education companies of tomorrow may well be as large as the pharmaceutical and healthcare companies of today.

Little wonder then, that there is a great deal of entrepreneurial interest in the education sector. One area that dates back many decades is coaching. Beginning with 'tuitions' provided by an individual to one student or a group, this evolved to coaching classes on a large scale—involving many teachers, many subjects and, increasingly, multiple locations. A further expansion has been into training on a scale where some of those involved even established institutions and built a brand. This facilitated the creation of a franchise model, making it possible to expand on a large scale and to do so rapidly. Probably the best-known examples are from the areas of IT training, where organizations like the National Institute of Information Technology (NIIT) and Aptech, provided trained human resources on the scale required by an industry expanding at a superfast rate. Some of them became global brands (NIIT, for example, set up training centres from Colombia to China).

This requirement for human resources in the IT sector also triggered a huge boom in private engineering colleges in states where they were permitted. Today, a vast majority of engineering students—especially in IT and related fields—study in private institutions. Also, there are now a few 'broad spectrum' private universities that cover a host of disciplines. A few also focus on the humanities. While some are genuinely philanthropic, many aim to be not only self-sufficient in funding, but to actually make a surplus (profits).

Privatization is, by no means, limited to higher education. Some entrepreneurs saw a big opportunity in preschool education (including day-care centres for very young children, which was especially useful to nuclear families, where both parents have jobs) and set up such places with an eye on large returns. With the right locations and good facilities, they could charge exorbitant fees. Similarly, sensing growing opportunities, some have set up 'international' schools (generally covering the curriculum for the International Baccalaureate [IB] examination/certification) or separate 'IB sections' in existing schools. Given the clientele that this is aimed at (those intending to send their children abroad for undergraduate education), they are able to charge huge fees.

Preschools/play schools and international/IB schools cater to a niche audience that has both the ability and the willingness to pay large fees. This has, therefore, become an attractive area of business. However, even in mainstream school education, there has been rapid growth in privatization. Here, the major drivers were the perception of poor quality of government (public) schools and the desire to have children learn English (hence the big growth in 'English medium' schools, even in rural India). As a result, the enrolment in government schools (across twenty states) fell by 13 million in the five years from 2010–11 to 2015–16.[26] In the same period, 17.5 million students were added to private schools. As a result, according to data from the District Information System of Education (DISE), about 35 per cent of children are now in private schools, compared with just 2 per cent in the '80s.[27] It is not clear whether one party in this competition has basically abdicated and given a walkover the other, or the other party is genuinely so much better that it is preferred, despite its generally higher fees.

Privatization has, by and large, taken place in areas linked to education or training for technology (engineering colleges, training institutions, or schools that emphasize computer education). However, it also means the greater use of technology: some kind of computer education to attract students (this, like 'English medium', is not only a magnet but an excuse for charging higher fees), and especially in training/coaching centres, the use of distance learning technologies.

The possible movement of education from the guru-shishya mode to corporatized entities is exemplified by Byju's, an educational technology and online tutoring firm. Founded in 2011 by Byju Raveendran, it began as one person tutoring students, and its success led to the creation of a global-scale company. In 2018, Byju's is reported as having over 15 million users overall and about 900,000 paid subscribers on a yearly basis, with an annual retention rate of about 85 per cent.[28] Their flagship product, available since August 2015, is a smartphone app named 'BYJU'S: The Learning App', with educational content for school students from Class IV to XII. The company also trains students for various competitive examinations in India, including for admissions to the Indian Institutes of Technology (IITs), Indian Institutes of Management (IIMs), the administrative services, and also for global admission tests like Graduate Record Examinations (GRE) and Graduate Management Admission Test (GMAT). The commercial success of Byju's is evidenced by the fact that it has raised funds of around $240 million since its inception.[29] In April 2018, its valuation made it a unicorn[*]. Apart from its entrepreneurial success, Byju's is indicative of the growing and large-scale use of technology (an app, in this case) for education.

A different technological scenario—a revolutionary, rather than evolutionary one—may be based on a much more extensive use of newer and emerging technologies. Thus, one could envisage a memory and learning chip implanted into humans, which would not only provide existing information and knowledge on a subject, but ensure continuous learning based on AI. Updates would be automatically provided through 24×7 online connectivity, ensuring zero obsolescence of knowledge. This would obviate the need to spend any time and effort on learning, and is clearly analogous to the machine learning that is already in vogue, where a machine is provided with all the required information to do a task and it keeps bettering itself through continuous, automatic self-learning. Robots, powered by AI, are already beginning to do this. Encapsulating this in a chip, implanting it in humans, moving from simple tasks to complex

---

[*]Named after the mythical beast, a 'unicorn' in the start-up industry is a company that reaches a $1 billion market value as determined by private or public investment.

ones, and onwards to a whole repertoire of knowledge—surely, this is not too far-fetched, given existing technological capabilities. I do not doubt that this will become possible five to ten years from now.

The question, therefore, is not if such a scenario is possible, but whether it is desirable. It is likely that some aspects of the 'softer' side of humans may be difficult to replicate by a machine or chip. Emotions and behaviours, often irrational and unpredictable, may be difficult for a machine to copy (think of Spock in the *Star Trek* movies and his inability to emote), but technology may well be able to emulate that too, through sophisticated AI systems and advanced machine learning. In any case, in this scenario, we are looking at the possibility of chips in humans and not complete robots. Over time, of course, this differentiation may become increasingly difficult.

Without delving into this, and its many philosophical ramifications, one can see that any such development would radically alter our concept of learning and education. Schools and colleges—whether in physical or electronic/virtual form—will no longer be required. 'Teachers' would be those who put knowledge onto a chip—and that, too, may get increasingly automated. In a simplistic sense, every child could be 'given' a PhD before s/he is even of kindergarten age!

The current developments in electronics, AI and neurosciences are definite indicators that this is not an impossible scenario. The question is not *whether* this will happen, but *when*. This scenario anticipates, like blended learning, a man-machine combine, in which technology will supplement human capability.

On the other hand, there are those who fear that machine learning and AI will not just supplement humans but, may, in effect, supplant humans. Machines that can out-think and outsmart humans may take over and run the world. As pointed out earlier, this frightening possibility has been articulated not just by a group of scaremongers, but by highly respected experts like the Nobel laureate (late) Stephen Hawking, and Elon Musk, amongst others.

One thing, though, is clear: technology is driving major changes in education, in all its facets—when, where and how it is delivered; by whom (human or machine) and from where; in what form; in assessments and corrective action; and in how it is administered. In years to come, we are likely to see even more drastic changes, with the 'classroom' of tomorrow as different from today's smart classrooms as the latter is from the under-the-tree gurukul.

As education is increasingly delivered through and by machines, learners of the future are likely to consider machines, rather than humans, as their teachers. One day, then, the popular song, 'To Sir, with Love', from the eponymous 1967 movie of the same name, may well be modified to 'To Silicon Sir, with Love'!

We began by outlining how economic growth is spurred by technology and knowledge, which draw from education. The contribution of education to technology has been growing fast, as the latter reaches ever higher levels of sophistication. On the other hand, technology is facilitating, accelerating and changing education in an unprecedented manner. The interaction between these two—technology and education—promises much benefit and gain, if we can ensure that it is a circular and upward-spiral relationship. Early signs of this are visible in the exciting new work across a wide range of fields, emerging from research centres and universities around the world. This will generate even greater interest and spur investment in technology-in-education. Clearly, education is on the cusp of radical changes.

4

ABC AND TECHNOLOGY

*Interaction Between Technology and
Attitudes, Behaviours and Cultures*

GLOBAL POSITIONING SYSTEM: NAVIGATING TOWARDS THE TRUTH

Driving to the airport the other day, I heard the Global Positioning System (GPS) on the mobile handset direct the taxi driver: 'In 600 metres, turn left to join National Highway 148A.' There was no reason for me to be surprised. Being familiar with the route, I knew the instruction was correct; nothing was amiss. Yet, I should have been astounded. After all, till only a few years ago, such non-human guidance from my home to the airport was unimaginable. Yes, I could follow the signboards if I had to go to a large, landmark facility like the airport; but what if the destination was a friend's home? Signboards would not exist, and I would have to depend upon my own memory and navigational skills, or on the directions given by roadside vendors and pedestrians. Now, of course, I just feed in the address and depend on my handset to guide me. In only a few years, such machine-generated directions have become commonplace and even people of my generation use this technology routinely.

A moment's reflection, though, brings home the impact of this phenomenal change. Earlier, navigating one's way to an unknown location was no easy task. When I first came to Delhi in the early '90s, an invitation to dinner was welcome, but also hazardous. The difficulty lay in finding an address in unfamiliar territory, made more challenging by the sometimes strange positioning of blocks in various Delhi colonies (for example, my home was in Q block, and R was logically adjacent; however, while F was just across, P was a kilometre away, to the other side of one of the city's biggest roads). On winter evenings, there was hardly a soul on the roads of residential colonies from whom one could seek directions. The search was based only on what I had been told earlier. In those pre-mobile phone days, one could not call and request guidance from the host while on the road. In early years, I inevitably got lost and finally reached the destination a long while after the indicated time. It is another matter, however—and as I soon discovered—that my 'late' was early by Delhi convention, where it seems a standard practice to reach at least an hour after the time mentioned!

Now all I need to do is enter the address and precise directions will be conveyed by the mobile phone, thanks to Google Maps and GPS. Not only that, I can also get a fairly accurate estimate of travel time, enabling me to plan an on-time (or fashionably late) arrival. I can also see where there is traffic congestion, which alternative routes can be taken, and the time differential as compared to the quickest route. What a change in such a short period of time!

As an aside, I must note that this technology does have a downside. At a recent meeting, one of the participants was late. Being a key player in the deliberations, the meeting could not begin in his absence. Embarrassed at holding up things, he came in flustered and quickly apologized, 'Sorry, I got stuck in heavy traffic on Ring Road, and it took me an hour to come from my office in Nehru Place.' At this, one of the others promptly said, 'But I checked on my phone; it said traffic is less than normal and the driving time from Nehru Place is just thirty minutes.' So, watch which excuse you give next time!

Similarly, in the good old days of (unreliable) landline telephony, you could paper over a forgotten call by saying, 'I tried to call, but just couldn't get through' or 'Your phone seemed to ring, but I got no response'. Such excuses were plausible, and also believed. Today, even a call you missed will show up on your records. Worse, social media platforms like WhatsApp tell the sender whether you actually read (or at least received) the message. Despite our famed system-beating innovative abilities, I am yet to hear of any good excuses that will help overcome the irrefutable evidence about traffic, calls actually made, or unread messages. So, one might conclude that technology is driving us to a truly seminal change: a new value system in which one has to always tell the truth!

INEVITABLY INTERTWINED: TECHNOLOGY
AND CULTURE

In most cases, technology was first developed to only substitute an existing task, so as to make it easier, quicker and/or cheaper. For example, it helps one to find the route to a friend's home more easily. Thus, it would seem that technology is a tool that is used within the existing framework of our lives. However, as we look around and back in time, it is clear that technology has always been a major influence on the way humans work, relax and live. Be it flint stones and fire, electricity, automobiles, computers or cell phones, technology has shaped the course of our lives in innumerable ways, with profound effects on our ABCs.

One seemingly trivial but telling example of how technology affects culture is from the world of telecommunications. It is a marked trait of some people to say 'hello' after every few sentences during a phone conversation, and to repeat themselves: 'Hello... How are you doing? Hello... I called to tell you that I am going to Kolkata. Hello... Hello... As I said, I am going to Kolkata tomorrow morning. Hello... Yes. Hello... Yes, I will call you when I am back...'

This characteristic is particularly pronounced amongst older people. In fact, a young friend of mine had also noticed this, and recently asked me why people of my generation kept saying 'hello' every few seconds, while conversing over the phone. This first struck me in the '70s, while travelling to the US. Phone callers there began with a 'hi' or 'hello' only at the start of a conversation, and often the caller spoke continuously for a minute or two before listening to the response. This difference in calling behaviour, I realized, was not so much due to a cultural chasm, as due to technology. In the US, the known and tried reliability of the telephone network enabled a caller to go through even a long monologue with the full confidence that the person at the other end was connected and able to hear him/her. In contrast, given the frailty of the network in India at that time, with calls often getting cut off mid-conversation, it was necessary to keep checking that the person at the other end was still on the line and able to hear. Hence, the frequent 'hellos'—an equivalent to what engineers call a 'handshake protocol' to ensure that there is connectivity. Over time, this method of ascertaining if the person at the other end was still there became a habit. Even as the phone system became more dependable, those who had lived through the years of an unreliable system continue, through force of habit, to say 'hello' after every sentence. Phone behaviour and the pattern of interspersing a call with 'hellos' got entrenched in the culture, thanks to technology.

In general, the technology-culture interaction is substantially asymmetrical, with technology affecting culture more than the other way round. There are examples, though, of culture impinging on technology. Here are two interesting ones. The first dates back to the '70s: at that time, Sunset, India's first drive-in cinema, was being constructed in Ahmedabad, where I was then living, and there was much excitement about this new 'technology'. Those familiar with such drive-in cinemas elsewhere in the world will recall that, at the time, they were mainly a refuge for young people seeking to go for an outing and yet wanting privacy. I remember going to some in the US and finding them full of cars, with not a soul in sight; rather like a ghostly parking lot. Of course, each car did have occupants—almost always two young people. All the windows of the cars were completely rolled up, except for one which was open only to the extent of being able to hang a wired speaker from it. The couples in each car were there to enjoy both the movie and an hour or two of complete privacy.

In sharp contrast, when the drive-in opened in Ahmedabad, it became a carnival. Each car came packed with people (young, very young and old), food and furniture. Generally, family and friends came in multiple cars, making it a group outing. Cars were loaded with durries or even folding chairs, along with food and drinks. During the movie—unlike the desolate parking lot in the US—hundreds of people sat under open skies, enjoying the movie, the camaraderie and the home-made food from tiffin carriers. Hardly anyone sat in their cars, and it was treated as a picnic with an accompanying movie. I am hard-pressed to think of any other examples of such a diametrically opposite use of 'technology', and how culture (group outings versus private outings) affected the way technology was used.

The second example, too, illustrates the strong influence that culture exerts, and is once again about communication technology. Today, India is known for the extremely low tariffs for mobile telephony, which are amongst the cheapest in the world. However, in the early years, calls were very expensive. In this scenario, those who owned mobile phones were obviously from the upper echelons of society. Yet, even here, the deep-rooted Indian cultural ethos of seeking (or extracting) value for money, predominated. It manifested itself in the form of that unique Indian innovation—the 'missed call'. Clever minds figured out that the mobile phone could be used to communicate an immediate message without incurring the high cost of a call (or even the lower cost of a text message). Thus, the missed call was used for everything, depending upon the context—from announcing one's arrival at a rendezvous point or destination, to letting your driver in the parking lot know that you needed to be picked up from the hotel porch. The receiver of the missed call, in the given situation, was aware of the likely message and did not need to take the call. So, communication took place at zero cost—an example of culture (in this case, the importance placed on 'value for money') exploiting and bending technology.

HUMANIZATION OF MACHINES

Culture varies across India, with strong regional and sub-regional differences in many aspects—language, dress, food, customs and rituals. Yet, there are many commonalities. One amongst them is the general garrulity of Indians. We have always had a strong oral culture, and are a vociferous and articulate people who love to talk and argue. Now, apart from 1.3 billion Indians, we have others to listen to and speak with: these comprise the growing number of machines, beyond the mobile phone, that have a 'voice'. Many new models of cars, for example, give audio warnings. The 'beep beep' of a reversing car has, in some cases, been replaced by a voice that says, 'This car is backing up'.

Machines 'talking' to each other is now a standard element of IoT, as data from sensors and equipment is passed on to other machines for processing or undertaking an action. As these machine-to-machine capabilities develop, they will 'talk' even more—possibly even competing with Indians! Meanwhile, and more pertinently, in the context of culture, evolution is taking place with respect to new technologies and AI. Now, we can speak to machines that understand instructions and act accordingly. You can ask for the lights at your home to be dimmed, the air conditioning temperature, lowered, or a particular piece of music to be played. You can find out the weather in any city, the score of the latest cricket match, or the latest news. A device—such as those powered by AI 'assistants' like Alexa or Siri* —costing just a few thousand rupees and connected to the Internet, can do all this and more. Researchers have been able to take this forward to the point where it is not possible to discern whether the voice at the other end is coming from a human or a machine. To make the difference undetectable, they have added the pauses and the 'umm' and 'aah' of typical human speech.

There are also long-term and philosophical implications of this 'humanization' of machines. In the near term, one can see the convenience effect: for example, bedridden or wheelchair-bound persons could benefit greatly. Those living alone can have surrogate company through a machine they can converse with. As someone who was living alone told me: 'With Alexa, I now have a companion staying with me.' Will the possible diminution

---

*Alexa is the assistant powered by Amazon's AI software. Siri is Apple's intelligent personal assistant.

in feelings of loneliness lead to more single-person households? The advent of TV and video games had already lowered human interaction; will intelligent, quasi-human machines lead to even lesser conversations? Will the need for human companionship decrease? We do not have all the answers right now, but as intelligent robots proliferate, they are bound to impact social life and culture in ways that are not yet very clear.

IN THE 'NEW TECH' AGE

We are already seeing the impact of technology on social life and behaviour. The spread of TV made a big difference to what children did during non-school hours. By all accounts, outdoor activities and interactions with other children decreased substantially. Added to this, was the subsequent proliferation of video games, which continue to have children hooked to their devices and act as a substitute for games played outdoors. More recently, broadcast TV is being supplemented by online content that can also be downloaded and viewed at one's convenience. Such online data streaming services, like YouTube and Netflix, have become very popular. Indicative of this is the hugely increasing data consumption: a senior executive at one of India's largest telecom companies says that 'overall consumption of data has gone up by eight to ten times over the last year'.[30] A study by Limelight Networks also says that Indians spend around five hours every week watching videos on the Internet.[31]

Research shows that TV addiction may have long-term negative effects on health: a study demonstrated a link between a sedentary, screen-heavy lifestyle amongst young adults (eighteen to thirty years) and cognitive decline twenty-five years later.[32] Another study linked binge-watching to loneliness and depression.[33] Experts opine that obesity in India is on the rise due to sedentary lifestyles and unhealthy diets.[34] In part, this could also be due to the content that children view—especially the promotion of unhealthy fast food. The reduced (non-school) interaction with other children probably has its own effects. Social skills and overall development depend on such interaction, and the lesser time spent on this is likely to have a long-term impact as well.

Late-night viewing of TV also means less sleep and, as a teacher told me, tired and less attentive children in the classroom. Another teacher gave me a different insight into the impact on children's health. She told me that because parents often watched TV late into the night (especially mothers, who have little time to view TV before dinner), they wake up later. As a result, the morning is rushed and the lunch carried to school (generally prepared by mothers in the morning), is something that is rustled together. This haste means that sometimes ready-made or quick-to-make food, which is not always healthy, gets packed. The teacher told me that as she casually saw a few lunchboxes and investigated further, she noticed that there was a perceptible change from the food that they used to bring as children, and that the trend was widespread.

TV content has also made a big impact in smaller cities and villages, and on people from lower socio-economic groups. It has brought the outside world into the insides of homes, and those who may not have ventured beyond their district boundaries are now exposed to a far wider canvas. Rural and small town women, whose movement is constrained by social factors, see the wonders of new places for the first time. Not only the sights, but also the clothes, behaviours, values and cultures of these places—whether in India or abroad—are viewed in homes across India. Undoubtedly, this creates new aspirations, as do the near-fantasy lifestyles of people in soap operas. New role models emerge, and their behaviours, clothes and lifestyles become something to be emulated. Many companies—especially those that sell clothes and cosmetics—seek to capitalize on this by reinforcement through advertising.

Evidence of the impact of this is the booming market for cosmetics around the country, including in smaller cities, towns and even rural areas. Another telling evidence is the proliferation of beauty parlours. Little wonder, then, that popular folklore attributes the spate of Indian women winning global beauty contests, through the '90s (and one in 2000), to commercial considerations. The theory is that the publicity that followed each such crowning made the winners role models, raising the 'beauty' aspirations of girls and young women in India. Although the last two decades have seen few Indian winners on the global stage, the boost provided to cosmetic products from the '90s has probably helped the rapid growth of a huge market in India. The implied reason for the victories is certainly unfair to the very deserving winners, but the impact on the 'beauty market' is undeniable.

On the sartorial front, the astute observation by a journalist-friend conveys well the change that has occurred. He told me that while covering the mass protests that followed the Nirbhaya incident in Delhi in 2012, one sidelight struck him: he noticed that a large proportion of the youngsters at the protests seemed to be similarly clad—in jeans. This was despite the fact that there was a huge difference in their social and economic background, with persons from upmarket, stylish and prosperous households cheek by jowl with those from conservative and far less well-to-do ones. By dress, it was near impossible to tell which youngster was from where. Another place to see a lot of youngsters, especially students, is in the Delhi Metro. Today, the similarity in attire is immediately visible—a pair of jeans and a T-shirt are standard. How much of this commonality is a result of exposure to TV is difficult to say, but the change is striking—from the days when sartorial style was a clear indicator of one's socio-cultural background, to today's near homogeneity.

This may reflect change that is merely superficial, but there is also a deeper change that has been triggered by TV, and amplified by the mobile phone. This transformation is related to behaviour and, to some extent, even values. For those who belong to families/communities where 'Western' clothing is not commonly worn, frowned upon and/or often disallowed, donning a pair of jeans is not so much about style as it is about independence or rebellion. This is particularly so for girls and young women from highly patriarchal regions or families (mainly in North India). In such families, young women are sometimes also not allowed to own a mobile phone—whether smart or feature—as it is believed that the exposure to a wide variety of content may act as a negative influence. It is also feared that it may be used as a secret means of carrying out 'illicit' conversations with men. However, many cloistered young girls have broken free from these constraints and use the mobile phone as a tool of liberation—one that enables interactions away from prying eyes. The mobile phone has, undoubtedly, spurred such male-female interaction. Clearly, technology is influencing not only their dress and behaviour, but also their attitude.

Many of these changes have been boosted by the ability to download/stream content on a mobile phone. Thanks to competition generated by the very low rates first offered by Reliance Jio in India, data can now be downloaded at rock-bottom prices. This has now resulted in Indians being the biggest mobile data consumers in the world.[35] According to the Telecom Regulatory Authority of India (TRAI), in three years—from the end of 2014—the average monthly data consumption increased from 0.26 GB to over 4 GB per person.[36] This is more than a fifteen-fold increase! Driving this was the unbelievable drop in data prices: from an average of ₹269 per GB in 2014 to ₹19 in 2018, and even lower in some cases of bundled services. With about 1.2 billion mobile phones (as of 2018), of which a third are linked to the Internet[37], access to all kinds of downloadable content is now very easy. The number of phones with Internet access is expected to touch 600 million by 2020.[38]

Sharing of videos has become common. In rural areas, one can now see small groups of five to six people crowding around one person showing some particularly interesting video—ranging from a short clip to a movie—on his mobile handset. In cities, I have seen the watchmen outside homes in residential areas watching videos on their cell phones. With little to do during their duty hours, this (or listening to music on the phone) is definitely a welcome diversion for them. The same goes for drivers when they are not driving—watching videos on a cell phone is a good way of passing time while awaiting active duty.

These downloads provide viewers more exposure to the wider world. One would, therefore, naturally expect that it would lead to better appreciation of different cultures, regions, customs and ways of life. Unfortunately, these expectations seem to be belied. Despite extensive viewing on TV and cell phones, the tendency towards openness, understanding and progressive attitudes is not in sight. In fact, one can argue with some justification that, if anything, attitudes and mentalities have regressed. For this, too, the link is probably technology. A chilling example comes from a report about graphic mobile phone clips of gang rapes being sold in shops in Uttar Pradesh.[39] The clips—thirty-seconds to five-minutes long—are, according to newspaper reports, being sold in hundreds and thousands every day, for ₹50–150.

It is ironic that the very means by which one had hoped for a more fraternal and harmonious society—through social media, online access and easy communication—are the ones leading to fragmentation, polarization and narrow-mindedness. Adding to this, are the immensely negative effects of the widespread dissemination of selective and false news, including doctored video clips. A fake news 'forward' received via messaging apps, which reflects and concurs with one's viewpoint, is promptly re-forwarded to dozens of others on 'similar interest' groups, whose other members would usually have a similar viewpoint.

Thus, biases and stereotypes get reinforced—even multiplied—and fake news spreads like wildfire. The sad and terrible consequences of forwards facilitated by social media going viral, manifested itself in India for a few weeks in mid-2018. Fake stories of children being kidnapped by child-lifters led to innocent people being brutally beaten—often fatally—by lynch mobs in various rural areas of the country, from Assam to Maharashtra to Karnataka. In one such case, the photos of four young men (visiting a relative of one amongst them) were circulated on social media as child-lifters; as a result, they were accosted by a mob and mercilessly beaten.[40] Tragically, one of them died.

Growing concerns about the circulation of rumours and fake news prompted the government to take this up with popular messaging app, WhatsApp. As a result, the latter initiated some new measures. Its statement said, 'In India—where people forward more messages, photos, and videos than any other country in the world—we'll also test a lower limit of five chats at once and we'll remove the quick forward button next to media messages.'[41] Limiting forwards to five users (instead of 250) and other measures may or may not be effective, but clearly this is a serious problem. Numerous people have also been lynched since April 2017 due to rumours of cow smuggling. One report says that as of July 2018, rumours of child-lifting alone have claimed thirty-three lives[42]. And social media has played a major role in the quick and wide circulation of such hearsay.

People everywhere are likely to be influenced by fake news and inflammatory (again, often fake) videos. Yet, it may be unfair to only blame social media for what follows; after all, lynching someone is many steps further from merely believing rumours from a forward. Such incidents indicate societal failure, and fingers may be pointed towards various institutions and factors. It is moot as to whether technology (social media, in particular) should be amongst the 'accused', though there is little doubt that, especially in its online and social media forms, it is substantially influencing attitudes and behaviours.

THE CATALYST

As noted earlier, in some areas, families do not allow girls and young women to own mobile phones, fearing the 'corrupting' influence the use of these devices could have on their social behaviours and morals. On the other hand, many families encourage—and even insist—that their daughters own a mobile, so that they can stay in touch if the women are out late in the evening. As a corollary, it has been an important catalyst in increasing the number of working women: families that hesitated in permitting women to take up jobs due to concerns about returning home late, are now letting them do so. The fact that they are accessible on the phone at any time (even during the journey) has given the elders a sense of comfort. This matters not only in the context of permitting them to take up a job, but also to go out at night. Between this and the overall change in the social scenario, the number of young women being out at night—be it for parties and dinners, or due to working late—has considerably increased. Again, one can see the role of technology in influencing the attitude of the elders, as also the behaviour and lifestyle of young women.

'Anywhere, anytime' connectivity via mobile phones may have made it easier for women to get parental and household 'approval' to work, but 'appropriate' jobs might also have been an issue. In this, too, it is technology that has been the key facilitator, in a broader sense. The boom in IT, the spread and reliability of global communication links, and innovative business models: these have converged and been exploited by smart entrepreneurs and companies to transform India into a global IT hub. With an insatiable appetite for qualified talent, this industry—barring occasional slowdowns—has actively and desperately scouted for suitable employees. As more women opted for higher education, including in the fields of software and engineering, they became a good source for such talent. Given the competition for human resources, each company tried to do something extra to attract women. Pick up and drop from home became standard for late night or early morning shifts (this became mandated by law, too), as much of the industry worked on a 24×7 schedule. This, and other facilities, added to the lure of a well-paid, white-collar job, making it attractive for young women and acceptable to parents. Concerns about daughters working the night shift were generally allayed, not only by the pick up and drop facility, but also by the always-on connectivity provided by the mobile phone. It is quite a sight to see thousands of young women working at night, in places like Gurugram and Bengaluru, day after day, in business process management (BPM) companies, which evolved from mere 'call centres'. Few things reflect the impact of technology on society as strongly and visibly as the phenomenal change in the gender ratio of the workforce in the organized industry. With over a third of its 3.9 million (in 2017) workforce being women[43], the IT/BPM industry has truly revolutionized the attitude to women and their work.

Interestingly, it has also triggered a deeper and more important social change. With the comparatively higher salaries in the IT/BPM industry as compared to other jobs, young women working there are often the highest earners in their households. While working and earning have, in themselves, given them a sense of independence, this added factor of earning more than the others in the family has had its own impact. Many argue, based on first-hand experience and anecdotes, that this has changed the dynamics of relationships in the family. If so, this can be considered a seminal change in traditional family equations in a predominantly patriarchal society. Technology, by creating these jobs (high-paying ones at that) and ensuring supportive conditions (phone connectivity, in particular) for women to work, has been the catalyst for major societal change.

EXPECT THE UNINTENDED

Mobile handsets are, inappropriately and somewhat archaically, still generally referred to as 'phones', connoting an instrument to make calls. In fact, over the years, this 'phone' has integrated within it, a vast array of functions; thus, making calls has become only one of its many uses. A quick look at the advertisements for handsets conveys the relative importance of its many functions.

The quality of the screen display, the features of its camera (or, cameras—in the plural—with one at the back and one in front), the storage capacity and the speed are the competitive USPs. None of these have anything to do with its original/basic function—to make and receive calls. A major use now is its camera function, used to record video clips or click photographs. In fact, mobiles that have a camera with 4K capability are even being used by a major Indian TV network to record and transmit videos.

One sees vast amounts of photographs and videos on social media, taken and circulated by people eager to share where they are and what they are doing (even pictures of food they are eating)—practically all the time. This includes the growing trend of taking 'selfies' and sharing them through social media. In many public spaces, particularly shopping malls, one can see dozens of people using their handsets to click selfies. I am astounded by the sheer number of youngsters doing this. All of them seem quite comfortable preening and posing in public as they take their own picture, looking at the result immediately, and doing numerous 'retakes'. The best of these are then promptly shared with friends through social media. It is not just in malls or restaurants, though: every flight I have travelled in, in the recent past, has had at least a few people taking selfies inside the plane.

This otherwise innocuous passion may well reflect a new behavioural trait—that of narcissism. On the positive side, it could mean a sense of self-confidence, comfort and pride in how one looks. The negative interpretation could be that it is a manifestation of self-centredness and an 'I, me, myself' approach.

However one views it, this change can be attributed as being facilitated by technology. The ease of taking photographs and sharing them with friends is made possible by the ever handy handset; therefore, looking at technology as a causal factor for this new attitude and behaviour is not all that far-fetched.

One unfortunate effect of the popularity—though 'craze' may be a more appropriate word—of selfies is the competitive pressure to photograph oneself in exceptional locations or unique situations. Thus, one sees people standing on rocks at the seashore with waves crashing into them, or at the edge of a mountain. I have even seen a person standing on a rail track and taking a selfie as a train approaches on the same track! Not everyone may be as smart and agile as that young man who jumped off the track in time to avert possible death. The eagerness to take an exceptional photo often makes people careless, leading to accidents. Every now and again, one reads reports of serious injuries—even death—where people were so engrossed in getting that perfect selfie that they neglected the most elementary precautions. Even as this is being written, in June 2018, I see a newspaper headline that highlights this danger: 'City woman falls off Matheran cliff while taking selfie'.

As in many other cases, here too, the problems arising from the unintended consequences of this technology have a solution that comes from technology itself. Researchers at the Indraprastha Institute of Information Technology (IIIT) in Delhi have devised an app for mobile phones that warns people about dangerous places for selfies. It does this through data analytics that identifies—based on past, reported incidents—dangerous locations and their coordinates.[44] If a person tries to take a selfie at the same spot, his/her phone identifies it through the GPS function and gives a warning of danger.

A somewhat similar and more widespread danger arises from the use of earphones to listen to audio or take calls. It is not uncommon to see people engrossed in whatever they are listening to, while driving or crossing roads. Quite oblivious to their surroundings, they are unable to hear the warning horns of approaching vehicles. This often results in accidents, as frequently reported in the media.

The sharing of photographs—not only selfies, but also of where one went and what one bought/ate—also reflects the changing attitudes towards privacy. It is now also common to hear someone speak over the phone, in public places, about matters once considered private. The voice is loud enough to not require eavesdropping; in fact, even if one is many metres away, they are forced to hear what is being said. In airport lounges, aircraft, restaurants and theatre lobbies, I have often had no choice but to hear things like, 'I want to tell you in advance that I am firing Rakesh Kapoor tomorrow, unless he sends his resignation today. What he did is not in keeping with the culture of XYZ.' (The actual name of the company, too, was mentioned!) Poor Mr Kapoor! A few dozen completely unconnected persons know the tragedy about to befall him. Here's another snippet, heard in a restaurant: 'What does that (two choice, unprintable abuses) think of himself? I will make sure that he doesn't get the contract.'

Like me, many have also often been subjected to someone's dinner menu. Here's a typical example: (as soon as the flight landed) 'I will be there in an hour. Make dal, bhindi and matar paneer, along with twelve rotis. And don't add too much green chilli.' In earlier times, people were more discreet about what they said—and how loudly—especially in public places where a large number of people are within earshot. Now, it's common to get inflicted by what is clearly a private conversation.

Can one blame technology for this? There is little data to draw any firm conclusion, but it is not illogical to assume that the sharing of one's every movement and moment has led to a redefinition of privacy. Certainly, norms have changed substantially with regard to what kind of conversation can be carried out in a voice loud enough for many others to hear. Is this because of the mobile phone and, therefore, another example of how technology has changed ABCs? Specific, evidence-driven answers may be some years in the coming. However, there is little doubt that technology already has had a considerable impact on our ABCs and the future is likely to see even greater change.

5

TECHNOLOGY, GOVERNANCE AND DEMOCRACY

How Technology is Changing Governance and Affecting Democracy

The day-to-day life of a common man inevitably includes—directly or indirectly—a number of interfaces with various authorities and their officials. These include municipal or local authorities, the Central and state governments, and their various offshoots. The transactions include the payment of utility bills (power, water, gas) and various taxes (income, goods and services, property); acquiring or renewing licences (driving, shop); seeking permissions (construction, staging an event); applying for or renewing official identity documents (passport, Aadhaar, PAN), and a number of other interactions. Till the '90s, these were almost fully dependent on hard copy documents, and many required a visit to the concerned office to submit the form and/or make a payment. However, over a period of time—and at a growing speed—these have been replaced by online operations. This transition to 'e-governance' has had many implications and these are best understood through concrete examples.

Till recently (and, in some cases, even today), land records were documented in physical registers in the office of a local authority (at the district, taluka or panchayat level). The sale of land, proving ownership for the purpose of loan, etc., meant that the land owner had to travel from his/her village to the office concerned, and hope that the authorized official (typically, the Talati in rural areas) was present, and not on leave or away for an extended lunch/tea break. The official then had to be met (not easy, given that a large number of people were vying for his attention) and he had to locate the appropriate register, based on which, ownership could be certified. Each stage in this process was fraught with uncertainty, beginning with the availability of the official. Not uncommonly, the first trip would be an abortive one, for one or the other reason (absence of the official, record not being immediately located, too many people in the queue, certificate can only be issued the next day, etc.), and a second or even third trip would be required. This meant that both transportation cost and time (which could have been used for other productive activities) were lost. Given the frustration and costs involved, many an applicant chose the 'classical' way of expediting the process—an appropriate 'incentive' for the official concerned.

The example of land records can be extended to a whole host of other areas where individual citizens have to deal with organized officialdom. Most officials are not corrupt, and the frustrations and delays are just part of archaic processes, made worse by overload and overwork. For the citizen, though, the experience is hardly pleasant and only reinforces the image of a powerful government machinery dealing with a cowering supplicant. This may well have suited the colonial masters who devised the system, or a feudal setup where subjects are expected to cower before an all-powerful king. However, it hardly befits a democracy.

PROCESS RE-ENGINEERING

Thanks to technology, change is now taking place. Land records have been digitized in most states. This not only makes easy the job of searching for a specific record, but can facilitate such access from anywhere, as long as there is electronic connectivity. Such access means that any official, located anywhere, can—if so authorized—issue ownership certificates or validate the sale and transfer of land. Some progressive states have enabled direct access by banks; so even a certificate is not needed in such states. Thus, for land or property dealings, it is not essential for people to go only to the office of the taluka/district concerned, but they could, in theory, go to a government land records office in any location. This effectively takes away the monopoly of the local office. Eliminating such monopolies of power is one way of reducing, if not stopping, corruption.

Digitization of land records has other advantages too. It is well known that a lot of crime and violence in rural areas can be traced to disputes centred on land, and many of these are caused by the untraceability or fudging of land records. Digitized data, duly authenticated and well stored, can bring certainty and transparency to records, thus reducing the potential for conflict. Such disputes can be further reduced by digitization of map data and correlating this with the position on the ground.

Pioneering work using IT (particularly in Andhra Pradesh and Karnataka), done almost two decades ago, has now spread to many other states, with digitization of land records being an example of the transition from conventional governance to e-governance. As can be seen, deriving the full benefits of this requires not merely digitization, but process re-engineering. In this case, it means changing procedures so as to provide access to the database by all authorized officials and giving multiple officials the power to issue the required certificates based on this. The digitization of land records was one of the projects under the National e-Governance Plan (NeGP), an ambitious effort to use technology and usher in e-governance in a number of vital areas in which citizens interface with the government.

Depersonalization, or the removal of human interface—which results from the introduction of technology into erstwhile human-to-human activities—has its own issues. Machines and technology are not very good at dealing with exceptions or ambiguity and anyone whose case is not within the standard pattern can find this frustrating. Also, many people prefer the personalization that comes with the human touch. However, the intermediation of technology is definitely advantageous and provides multiple benefits. These include:

- Instant access, from anywhere, to applications or forms; these can also, in many cases, be filled up online.

- Online submission of forms as well as copies of any certificates or documents that may be required (by scanning and uploading).
- Data on applications (date, time, etc.), enabling the creation of a dashboard for monitoring (and corrective action) by higher authorities; also facilitating periodic review of changes required in the process or policy.
- Tracking of the progress of his/her submission through various stages of the process, by the applicant. (Of course, this requires processing to be done entirely digitally, with the status being accessible online.)
- Online issue of certificates, receipts, etc., enabling printing, if required, at the customer's premises.

These could obviate, or minimize, the need for travel, thus saving time and money for the applicant. They could also speed up the overall process. Depersonalization reduces the scope for corruption, as well as the unnecessary wielding of discretionary powers.

The advantages of technology and the resulting disintermediation are clearly visible in the case of DBT. Payment of wages for initiatives like the Mahatma Gandhi National Rural Employment Guarantee Scheme (MGNREGS), disbursal of pensions and scholarships, assistance for housing (under the Prime Minister's rural housing scheme)—all these and more are now paid directly from a central point (mainly from the Central government) into the bank account of the beneficiary. This has cut off a number of intermediaries who were part of the process earlier. As a result, disbursal of cash is much faster and the possibility of corruption has been vastly reduced. Further efficiencies have been ensured by automation at the backend, in the treasury operations. Of course, all this requires robust IT systems, backup data centres and strong cybersecurity measures. These are ongoing and ever increasing requirements, but a beginning has already been made.

As a further use of technological opportunities, the Ministry of Rural Development has leveraged geographical information systems (GIS) in its projects for housing the poor, referred to as the Pradhan Mantri Awas Yojana (PMAY). In this, disbursal of payments is linked to the actual evidence—and not merely certification—of progress. This is done by taking a photograph (date-stamped) of each major stage of construction, with the location identified through GIS and the beneficiary identified through Aadhaar. This enables the release of funds for each location, based on the visual evidence of the progress.

The data resulting from all this is truly massive—hundreds of millions of transactions, extensive information on individuals, infrastructure, geographical locations and projects, and their progress for a few hundred thousand villages. This is, indeed, a rich storehouse and ideally suited for use by the new tools of data analytics to provide insight, generate reports, identify discrepancies and throw up alerts. All this can certainly help to improve service delivery, even as costs are contained or lowered.

One example of the use of this data is its analysis on the basis of the geographical demarcation of State Assembly and Parliamentary constituencies (different from the block and district boundaries that are of interest to officials). Creating summary dashboards of the status will help elected officials (members of the State Assembly and the Parliament) to better monitor the progress of various schemes in their constituency. It could also provide an impetus for improving the performance by creating, formally or informally, competition at different levels (panchayat, block or district), even as it helps to identify the best practices for possible emulation (with adaptation as necessary).

BIG CHANGES, VISIBLE IMPROVEMENTS

Many years ago, as part of the upgradation and digitization of its processes, the Ministry of Corporate Affairs (MCA) initiated its MCA21 project. This was a path-breaker in the use of technology in the government in India. It was based on a new and revolutionary concept: the procurement of services, and not of hardware and software. The vendor/partner was free to use the hardware and software of its choice, as long as it met the defined service requirement, including throughput, time, system downtime etc., as detailed. This converted what would have been capital expenditure by the government (in purchasing the equipment and software), into a yearly amount or operational expenditure. This model—of operational expenditure instead of capital expenditure—is, of course, widely used now, especially in the organized industry (airlines, for example, lease aircraft rather than buying them). For the government at that time, it was indeed a bold innovation, as this model takes care of the problem of equipment obsolescence. In government procedures, the process of procurement may take a year (or more, in some cases), rendering obsolete the specifications originally laid down, as newer capabilities emerge. Take computers, for example: in the time taken between laying down specifications and placing the order, completely new models with enhanced capabilities will be available in the market, sometimes even at a lower cost.

The other innovation in such business models is the basis of payment: a per-transaction fee, instead of a fixed yearly amount. This ensures that the vendor has a business interest in maximizing the number of transactions and so ensures that the process is easy, efficient, reliable and customer-friendly.

Some years later, this was further developed during the project of digitizing the process of issuing passports. A partner was selected and the full process of passport issuance was outsourced, barring sovereign functions like police verification and final approval. This brought about a sea change in the speed, efficiency and general quality of service. The private sector partner introduced automation, digitizing most of the steps involved. In addition to the extensive use of technology, things like modern offices, with comfortable waiting areas for applicants and due attention to customers' needs, made for a better user experience.

However, in speeding up the overall process of issuing passports, one bottleneck that remained was police verification. The standard steps involved district police headquarters (HQ) accessing the Passport Seva application (developed by the private sector partner) and downloading the police verification forms sent by the Passport Seva Kendra [PSK] (also run by the private sector partner). These forms were then printed and sent to the police station of the area where the applicant lived. A policeman would then visit the applicant's house for verification, collect identity documents and also get signatures from two references. Once cleared, the paper file would be sent back from the police station to the district police HQ, which would, in the final step, send it to the Regional Passport Officer. However efficiently done, this multistep process would take a long time. Even in the 'millennium city' Gurugram, it would take a few weeks just for the police verification.

Now, thanks to a technology-based process, the Gurugram police intend to reduce the time to two days.[45] First, a simple process change: the PSK directly assigns the police verification in the work list of the police station concerned. Then, a specially trained policeman, equipped with a tablet computer, does an on-the-spot verification and uploads the report (including a photo, electronic signature of the applicant and the identity documents, after verifying them) on the mPassport Police App of the Ministry of External Affairs (MEA). Apart from vastly speeding up verification, this has also made the process paperless. This recent change in procedure is indicative of how simple technologies (a tablet, in this case) combined with apps, software, connectivity and process re-engineering can bring about big improvements.

A similar approach of technology-enabled and customer-friendly solutions is being used for issuing driving licences in many cities, and has certainly made the process smoother and hassle-free. Touts and corruption may not have been altogether eliminated, but the reduction in both is very visible, and those who have undergone the process recently, will vouch for this. Earlier, whether one could or could not afford it, they had to inevitably go through an intermediary (for some, the well-connected representative of the Automobile Association; for most, a 'facilitator'). Now, as I have also experienced, this is no longer a necessity. One can seek an appointment online, go to the Regional Transport Office (RTO) and, with a minimal waiting period beyond the appointed time, get the whole process done easily, without any special help.

EASE OF LIVING

The global index measuring the ease of doing business is considered an important factor in determining the extent of foreign investment in a country, besides speeding up the transition from intent to output. The 2018 ranking put India at #100, a big leap from its previous year's rank of 130.[46] The government has also set a target of being in the top fifty. While this is of obvious importance, to the common person, it is the technology-driven changes in the citizen–government interaction, particularly on seemingly small and unimportant matters, that make a big difference.

This facilitation in day-to-day living is not only critical at the individual level, but also contributes to the overall efficiency of the country. The ease of living is, therefore, as important from the economic point of view as it is from the societal angle. Technology is beginning to make a visible difference in this area, with the advent of newer technologies and in combination with innovative business models. New apps make it possible to access one's utility bills on a mobile phone and immediately pay them through a mobile wallet loaded in the phone or from one's bank account through a Unified Payments Interface (UPI). Smart cards allow the use of multiple modes of public transport, with the automatic charging of fares eliminating the need to buy tickets/tokens for each journey. Tracking of trains and updates on their expected arrival/departure times is now common. This is being extended to local bus services too. So much of governance is now online that visits to government offices are now a rarity for many people. It is these ease-of-living facilitators that truly make a difference to the common man. All this is indicative of the profound effect that technology is already having on governance; clearly, there is more to come.

Much of what has been discussed thus far, is at the micro level—relating to that of the individual—and the impact of technology on his/her interactions with the government. However, technology has also had a very big impact on the much broader, macro level of organizations and society.

POWER TO THE PEOPLE

Over three decades ago—and before the Internet—audio cassettes, photocopiers and scanners were dubbed as the 'technologies of freedom'. This followed their extensive use for propagating news of various protests and circulating views against totalitarian regimes— for example, in Iran and China. The technologies were used not only to bring together anti-regime groups within a country, but to convey information to and from the outside world. For example, in the '70s, rousing speeches by Ayatollah Khomeini (then in France) were circulated in Iran by making thousands of copies of audio cassettes, which was then a 'new' but easily accessible technology. Friends from Iran have told me that this was an important factor in galvanizing and mobilizing the people within their country, leading to the revolution in 1979. China, however, was a different story: while news from within the country brought much foreign support for the protesters, the regime was strong enough (and, many would say, ruthless enough) to quell the protests that had flared up and peaked in 1989.

These then-new technologies opened up new vistas, where content could be transmitted to tens of thousands of people quickly, easily and at a comparatively low cost, with little possibility of governments being able to control it. It was in this context that the Internet seemed an even more important tool of people's empowerment. The dawn of a new age of 'power to the people' seemed imminent.

In its early years, the Internet was seen as a force of empowerment and a means of correcting the power asymmetry between the individual and large organizations (including the government), which had taken place as these entities grew larger and wielded more clout. With the Internet, any individual could put out information and express a view, potentially reaching millions. Long regarded as a passive recipient of information and views put out by the government or media organizations, the individual suddenly became a source, capable of generating and transmitting whatever s/he wanted. Many hoped—even expected—that the Internet would redress the inequity and give more power to the individual. The ability to influence and shape opinion was, it seemed, no longer the monopoly of large media organizations or the government, but could also rest with the individual. Devolution and decentralization of power seemed inevitable and just around the corner.

However, that imagined future was not to be—at least, it has not happened so far. The proliferation of a multitude of sources made for an overload and led quickly to the evolution of content aggregators. These portals are gateways to content, but also gatekeepers. This makes the concept of a completely democratized Internet an unrealistic one.

In recent times, there has been a phenomenal growth of innovative platforms, some of which have leveraged the Internet's capability to connect people. Facebook and WhatsApp are two of the most well-known apps. While the former has 2.2 billion monthly active users globally[47] and 270 million users in India (Q2, 2018 figures)[48], the latter has 1.5 billion worldwide and 200 million in India (February 2018)[49]. Twitter, the platform for short messages (formerly 140 characters, which got upgraded to 280 for most languages), is another popular forum, with 335 million monthly active users in early 2018[50] (10 million in India[51]). The need for brevity in Twitter's short format, has led to the 'murder' of both grammar and spelling, converting English into a phonetic language. Often decried for these reasons—obviously, by older people bemoaning the younger generation's abandonment of the very foundations of language—it actually harks back to earlier times: the days of the telegram.

The telegraph, developed in the nineteenth century by

Samuel Morse and others, began with messages being conveyed instantaneously from point to point through wires, using equipment that generated (and received) electrical signals of only long and short 'beeps' (dots and dashes, respectively). These were translated from and to words on the basis of the Morse code. Thus, an SOS (emergency message/distress signal) was conveyed through three dots, three dashes and three dots (...- - -...). This Twitter-length explanation may be necessary, especially for millennials, since the electric telegraph is now extinct in India, about 160 years after the first public telegram was sent between Mumbai (then Bombay) and Pune in 1854. As a person who was, at one time, a regular user, I made it a point to go to the telegraph office in Delhi on 15 July 2013—the date which marked the closing of the telegraph services in India—and sent myself a telegram!

The charge for sending a telegram was based on the number of words, and so reducing the length of the message was a necessity. Words that were redundant to the core message were obviously eliminated. Punctuation had to be spelt out; so, in longer messages, unnecessary punctuation was also dropped (generally, the comma) and 'full stop' became just 'stop'. As an example, a message like 'I am reaching tomorrow, at 5:45 p.m., by Frontier Mail. Please arrange for a car to pick me up, and meet me at 10 a.m. on Wednesday, 16th', would be formulated in 'telegraph-ese' as 'Reaching tomorrow 1745 Frontier stop Arrange car stop Meet 1000 Wed'. This reduces the word length (including punctuation) to a third, while communicating the message clearly. Of course, this is not great English! Grammar was sacrificed for brevity. Ultimately, as the younger generation may want to remind pontificating elders, this is not too different from a Twitter message!

A FALSE DAWN?

Initially, there was hope that the exchange of content and views through the so-called 'social media' would enable dialogue by providing a platform for discussing differing viewpoints and perspectives. One envisaged lively debates, with various opinions contending for wider acceptance, almost like a 'civilized' college debating society. Reality, though, has turned out differently.

Increasingly, these new platforms and their ever increasing versatility are used for exchanging inanities ('Look at this ice cream that I am eating!') or passing on unverified reports. The result, as noted earlier, is the rapid propagation of rumours and falsehoods through the amplifying effect of 'forwards'. Anecdotal evidence indicates that the more sensational or outrageous the post, the faster and wider it gets circulated, with the impact multiplied by the echo chamber effect. This has encouraged the planting of fake news, which is capable of not only creating fear, enmity and hate, but even triggering riots. As a result, there is growing global concern over fake news.

Data mining and clever algorithms verging on AI enable platforms like Google to customize advertising. The ads that pop up when you log in are geared to your interests, as deciphered from your past online behaviour (what has been searched, which sites have been visited, how long one has remained on a particular site, etc.). Platforms also use this to customize news. Reading news—true or fake—that interests you and matches with your world view obviously adds to the echo chamber effect. There is also concern about trolling, abuse and threats on social media. However, rooting out fake news, rumours and abuse is not easy.

These developments, and the way social media tends to be used, have done little for the promotion of cross-cultural understanding, or reflection and dialogue based on varied perspectives. Instead, in many cases, they have strengthened sectarian identity, reinforced biases and sharpened societal divides. The hope that these technologies would democratize, empower and enlighten seems to have been thwarted. Maybe it was a false dawn, after all.

On the other hand, in all fairness, such a critique may be one-sided. Like many other innovations that greatly impact society, technology, too, is a double-edged sword. Common interest platforms, identity strengthening and the sharing of a viewpoint are also useful in mobilizing people and creating mass movements. As mentioned previously, in India, this power of social media was first strongly seen after the horrific Nirbhaya incident, when tens of thousands of people assembled for protest demonstrations, mobilized primarily through social media. The fact that this force is also used for negative purposes is not necessarily the fault of social media. Yet, there are concerns regarding the dangers of polarization and majoritarianism that social media engenders, and the effect of controversial content— whether true or doctored—on groups already susceptible to a certain viewpoint.

WHITHER DEMOCRACY?

It is possible that the context determines the impact of technology. In a climate of ultra-nationalism and heightened cultural and religious schisms, technology seems to aid alienation and a sense of 'otherness'. In a more benign context, it may well promote better communication and understanding. Is it then an amplifier, with the role of accelerating and widening the dominant tendency? It may be too early to reach a definite conclusion, which would require more data over a longer period.

In the present context of concerns and, sometimes, even paranoia about terrorism, some technologies have even led to the curbing of freedom. CCTV cameras, for example, are now proliferating rapidly across the globe. In some cities, every movement outside homes can now be monitored 24×7. Increasingly intrusive monitoring has become more acceptable, as governments across the world convince people that this is a necessary element of tracking and countering terrorists. In many countries, the tapping of phones and/or accessing of mails by the government are also becoming acceptable practice. Even in the bastion of democracy—the US—it is reported that all telephone calls are recorded.[52] The US is also known to have tapped the mobile phone of German Chancellor Angela Merkel.[53] The loss of privacy, and hence, of a basic right and freedom, is perceived as the price one has to pay for security.

Using the widespread CCTV network, new technologies of face recognition can quickly identify someone in a crowd. Databases, including data from analysis of social media as well as that from means of identification (such as Aadhaar in India, or the Social Security Number in the US) can then potentially be used to identify a person. Linkages of these with bank accounts, credit cards, cell phones, etc.—through data analytics—provide full details of a person's whereabouts, movements, financial transactions, expenses, lifestyle and much else. By tracking these, sophisticated algorithms, analytics and AI can model behaviour. Predictive analytics can use this to forecast how you will behave and what you will do in any given situation. If such predictive analytics indicates that you might perform a violent act, will you be subject to preventive arrest? In the past, this used to be common fodder for science fiction stories; it may now become commonplace in reality.

There have also been experiments using cameras inside planes to detect suspicious behaviour through a variety of visual inputs, including the monitoring of eye movement, dilation, etc. These could possibly help to identify a potential terrorist. In such a case, is pre-emptive action justified? Even though there is an error-probability in this model, in today's climate, a majority of people in many countries are likely to favour an anticipatory arrest.

The regular shutting down of Internet access by totalitarian regimes could be interpreted as a sign of the Internet's power to promote freedom (else, why would access be prevented?). Yet, the reality is that the same totalitarian governments use technology as a means of control and monitoring. In such cases, technology is clearly a powerful tool of oppression. Even in a democracy like India, shutting down Internet access is not uncommon, especially when serious trouble is anticipated. A report indicates that between May 2017 and April 2018, there were as many as eighty-two instances of Internet shutdown in India.[54] Further, the report notes: 'South Asia has witnessed the highest number of Internet shutdowns globally, with India earning the dubious crown for the country with the highest number of Internet shutdowns'. This curbing of freedom is done with the intent of maintaining law and order, thus making it out to be a measure to protect democracy, which provides yet another example of freedom being far from synonymous with democracy.

How should one react to the infringement of privacy, loss of anonymity and implicit curbing of freedom? Is this a consequence of technology, or of the socio-political context? There is probably no definitive answer. However, it does seem that the romantic idea of 'technologies of freedom' has been neutralized by the harsh truths of the real world. Ground realities do not offer evidence of new technology catalysing more freedom or contributing to a more democratic world. The only definite conclusion is that there is no definite conclusion. It seems that the apt analogy here would not be the curate's egg (good in parts), but the surgeon's knife (used to save lives, but can also be a murder weapon).

On a positive note, as one looks around the world, there can be little doubt that new technologies have enriched and improved lives. Despite many concerns, one cannot but be optimistic about their future role and contributions.

6

SOFT POWER: MIND VERSUS MUNITIONS

Could Weapons of Mass Communication Replace Weapons of Mass Destruction?

'Political power grows out of the barrel of a gun,' asserted Mao Zedong, once the supreme and unquestioned leader of China. Ninety years ago, when he first made this statement, power did largely depend on the force of arms, and countries that were militarily strong called the shots. After the Second World War, the two superpowers (the US and the Soviet Union) were defined in that manner, because of the size and strength of their military arsenals. Yet, even as both kept adding to their store of weapons, the battle between them was positioned as an ideological one—as a contest between two different ways of life. Generally speaking, the US stood for freedom, democracy and liberty, while the Soviet Union positioned itself as a champion of the oppressed, seeking equality and the common good for all people. Thus, even as both produced more weapons—each more deadly than the one before—the true battle was for the hearts and minds of people around the world.

Some years before Mao's pronouncement and the US–Soviet arms race, one man had taken on the seemingly impossible task of overthrowing colonial rule without the force of arms. Mahatma Gandhi demonstrated, through his principles of non-violence and Satyagraha, the power of 'no power'. Time and time again, through the decades of India's struggle for freedom, he showed how those with no weapons could also triumph over the well-armed, and how morality could trump armaments. Drawing inspiration from Gandhi, leaders like Martin Luther King, Jr. in the US and Nelson Mandela in South Africa demonstrated the superior power of morality further.

The battle between mind and munitions is an ongoing one. While history bookmarks wars, this deeper and long-running struggle is largely unwritten, with only a few of the prominent events being documented and discussed. While Mao succinctly articulated a largely unchallenged truth, a contrary perspective, too, has long been around—'the pen is mightier than the sword'. Military power, though, gives one a different perspective. In 1935, Joseph Stalin, the then all-powerful leader of the Soviet Union, was advised by a French politician that he should win the favour of the Pope, who could influence Catholics to counter the growing threat of Nazism. Stalin is reported to have scoffed: 'The Pope! How many divisions* does he have?' Yet, few will

---

*A division is a large military unit or formation, usually consisting of 10,000 to 20,000 soldiers.

doubt that the Pope was, in many ways, as powerful as Stalin. Today, too, the Pope wields exceptional power and influence (even without many divisions!).

A FORCE MULTIPLIER OF GREAT POTENTIAL

Despite many heralded military conquests, ideas and ideologies have, for centuries, continued to be a power to reckon with. Thanks to technology (primarily ICT), in the last few decades, the dissemination of ideas and the speed with which this has happened has vastly increased. New technologies—including satellite broadcasting and mobile telephony, along with the Internet and broadband—have created near-universal availability and low-cost access to communication. This has made possible the global flow of communication, with borders becoming less relevant. In the case of goods, boundaries matter since countries have regulations regarding checking and imposing entry tariffs on items that are imported—or even stopping them. The same principle also applies to the cross-border movement of people, through immigration checks and visa fees. In both these instances, countries are able to actually enforce these regulations. In the case of communication, though, no such regulation is feasible. There was a time when countries banned and blocked the entry of certain newspapers or books—a practicable proposition till they were available only in physical form. However, such material is now available in 'soft' (electronic) form too. To block these, some countries do try to limit access to certain websites and broadcast channels. However, people have inevitably—and quite easily—found ways of circumventing such restrictions. Historically, it has been difficult to restrict the flow of ideas; today, with the capabilities of communication technology, it is well-nigh impossible. The result of this widespread and virtually unlimited reach of communication is the amplification it provides to an idea or point of view, vastly increasing the latter's power. Communication is, thus, a force multiplier of great potential.

This power is also used by military forces to enhance their effectiveness, which began as C3I (communication, command, control and intelligence); the addition of computers made it C4I, often with S added for surveillance. Now, with the addition of 'cyber' as a mark of the growing importance of cyberwarfare, there are five Cs. But then, technology has always been an important element of warfare, and often the crucial differentiator that decides who the victor will be.

The use of technology in war, though, is going through a radical change as military forces across the world are beginning to realize the great importance and impact of the morale of the civilian population and the perceived morality of the cause. Of course, the concept is not new. Some two and a half millennia ago, circa 544–496 BCE, the Chinese general and military strategist Sun Tzu had said: 'The supreme art of war is to subdue the enemy without fighting'. Centuries later, Carl von Clausewitz (1780–1831), the Prussian general and influential military theorist, put it differently—'War is the continuation of politics by other means'[55]—implying that battle is but one of the means of asserting and ensuring the supremacy of one's view.

During the Battle of Britain in the Second World War,

England suffered bombing raids by the Germans daily and at any time of the day or night, forcing the civilian population to scurry to bomb shelters. Adding to the fear of bombs was the shortage of goods and the distinct possibility of a German invasion. In these difficult times, what kept up the morale of the people were the inspiring words of Prime Minister Winston Churchill, carried far and wide by the British Broadcasting Corporation (BBC): '…We shall fight on the beaches, we shall fight on the landing grounds, we shall fight in the fields and in the streets, we shall fight in the hills; we shall never surrender…' Even after the disastrous retreat of the British army from France in 1940 (captured so powerfully in the movie *Dunkirk*), this speech broadcast is remembered for instilling courage. Broadcasts by BBC and others were also instrumental in informing and sometimes also conveying coded instructions, while encouraging the 'resistance' in the occupied parts of France and other places in Europe.

At the same time, the German propaganda machine was equally active. If Churchill attained iconic status for powerful, inspiring speeches that charged up his countrymen, Paul Joseph Goebbels (the then German Minister for Propaganda) was infamous for propagating false information. 'Goebbelsian' has become a term to describe the dissemination of untruth, particularly by constant repetition of a falsehood. Of course, taking liberties with the truth was by no means his monopoly: the Allied forces arraigned against Germany had their own propaganda units that continuously churned out information—and sometimes even false information—to influence the Germans and the others. Thus, fake news was used for political gain long before US President Donald

Trump popularized it in contemporary discourse!

If the Battle of Britain indicated the importance of communication (and, obviously, leadership) in steeling the resolve of ordinary people, the siege of Leningrad (today known by its original name, Saint Petersburg) in Russia, was the zenith of people's determination being the most important factor in thwarting the enemy. For 872 days starting 8 September 1941, millions of Leningrad's inhabitants withstood the siege of their city by German forces. Surrounded and under continuous attack day after day, they refused to surrender, despite the depleting stock of food and other essentials. Once again, morale defeated might.

VICTORY WITHOUT WEAPONS

A quarter-century after Leningrad's victory, Vietnam provided many lessons about morals, morale and might. Without detracting even an ounce from the weighty win wrought by General Võ Nguyên Giáp and the Việt Cộng, over the might of the US, one must recognize the role of the peace movement in bringing global pressure on the US to withdraw from Vietnam. The '60s philosophy of 'make love, not war' did not quite conquer the world, but it certainly inspired young people everywhere. Within the US, the peace movement helped to tilt the balance, ultimately resulting in US capitulation in Vietnam. What gave added strength to those who favoured peace, was not just the demonstrations and mobilization on campuses, but also the media coverage of these. Communication technology, by enhancing the reach and effectiveness of the media, took the war and its horrors into people's homes in the US and around the world. It also amplified the impact of the peace movement. The nebulous concept of 'public opinion'—including global public opinion—suddenly became important. Media, in conjunction with communication technology, acted as a force multiplier and became a new weapon in the soft power arsenal.

Another two decades later, as China's military muscle moved in to crush demonstrations for greater freedom, the 'morals versus might' battle was captured by one powerful image: that of a lone individual blocking the path of an advancing line of battle tanks. As this photograph made its way from Tiananmen Square in Beijing to around the world (in near real time, even in 1989—thanks to communication technology), it created a tremendous impact. Even China's leadership, often impervious to global opinion, had to take note of the powerful reaction that this photograph triggered.

More recently, the world saw pictures of Syrian refugees fleeing their war-torn country, including an image of a body that had washed ashore—it was of a young child who was attempting to escape the horrors of war along with his family. These visuals unleashed a great wave of sympathy and the anti-immigrant mood saw a substantial change. This is another instance of how technology can influence people and affect policies.

These examples—from Sun Tzu to the Syrian refugee crisis—are indicative of the historical importance of non-military factors in deciding the outcome of battles. With its widespread, easy and low-cost access, communication has become a major element in influencing populations even before 'battles' take place. Doubtlessly, in some cases it has even made fighting unnecessary, with the power of arms being overwhelmed by peaceful mobilization. The 'colour revolutions' in the formerly Communist countries of Europe in the early years of this century, which were a fallout of the collapse of the Soviet Union, are an instance of the power of such peaceful upsurges.

There is another example: in the first years of the current decade, beginning with Tunisia in December 2010, the 'Arab Spring' saw totalitarian regimes in a number of North African and Arab countries being overthrown in a mainly peaceful manner. Interestingly, credit is generally given to social media, which served as the primary means of communication and mobilization. The fact that, in many cases, the promise of democracy was ultimately belied is a different matter. This does not take away from the demonstrated power of communication in ushering in radical change. Contrast this with the billions of dollars spent by the US—and worse, the thousands of lives lost in military action—in bringing about a 'regime change' in Iraq and Afghanistan, and attempting the same in Syria. The results, in all cases, have been disastrous. On the other hand, the use of communication (rumoured to be actively supported by the US) in Eastern Europe was far more successful in bringing about regime changes—that, too, peacefully.

Psychological warfare is, at least in some instances, more effective than physical warfare. Communication—powerful as it has proven to be—is but one tool in the armoury of peaceful 'weapons' that have often been more effective than military force.

These have been collectively christened 'soft power' by Joseph Nye in his book, *Bound To Lead*.[56] It is a means to influence or persuade, to shape preferences, without coercion, which—by contrast—uses (or threatens to use) force or hard power. Nye contrasted co-optive/soft power with hard/command power in another of his books.[57]

The most important part of soft power is that it is non-coercive. Nye identifies its key elements as 'culture', 'values' and 'policies'. He offers this definition: 'This soft power—getting others to want the outcomes that you want—co-opts people rather than coerces them.' Nye asserts that in the age of information, 'credibility is the scarcest resource,' underlining the value of credibility. Little wonder then, that countries around the world seek to 'create' credibility for their viewpoint.

While Nye can be credited with introducing the term 'soft power' into contemporary strategic and diplomatic discourse, the basic idea is millennia-old. One has only to go back to Sun Tzu, who recommended that the aim must be 'to subdue the enemy without fighting'.

In fact, for some time now, the irrelevance of defending or capturing territory has been obvious. The bigger battles have long been for people's minds. The recognition of this has brought the concept of soft power to the fore, and led countries to see how they might best use their special strengths in any particular facet of soft power.

For a few centuries, till the middle of the twentieth century, the West—especially the United Kingdom (UK)—dominated the world. Small European countries, like Portugal and the

Netherlands, had colonies many times their own size. With the end of colonial rule, they withdrew. However, they left behind a legacy of language and culture that, in many places, is still a powerful force. In India, English is the language of the elite, of the judiciary and much of the administration. It is also the passport to upward mobility—social, economic and geographic (from a village to a small town to a city). Western culture is strongly embedded in middle-class India, various institutions and organizational structures and processes; laws from colonial times are very much part of how India is governed. Our worldview, too, is greatly influenced by all this, as we tend to look at global issues through the lens of Western values and perspectives. The fact that this is true even today, seven decades after India's independence, speaks volumes about the power of culture and ideas—the soft power of Britain.

INDIA'S GROWING SOFT POWER

India, too, has long been an exporter of culture. From the first century CE, intrepid travellers, seafarers and traders have carried Hinduism to Southeast Asia. Today, despite an overwhelmingly large Muslim population in Indonesia and an overall 1.7 per cent Hindu population[*], Bali (the province with the largest Hindu population in Indonesia—83.5 per cent[**]) continues to treasure its Hindu temples, traditions and heritage, and stories from Hindu epics are regularly enacted.

---

[*]This figure is as of the 2010 census
[**]This figure is as of the 2010 census

On a visit to Indonesia, a tour guide regaled me with many a long and colourful story of Agastya, a much-respected sage of India, whom many refer to as the father of the Tamil language. A temple devoted to Agastya stands on a rock just off one of the beaches, and he is said to have lived there. Legend has it that there is—or was—fresh (potable) water always available within the rock, though it is surrounded by seawater. Agastya is remembered through art and sculpture in many temples in the country, as also in Cambodia and Vietnam.

Indian cultural influence is also evidenced in the names of Indonesians. Amongst the better known are those of three former presidents, whose names have a Sanskrit origin. The name of the first and longest-serving (1945–1967) president of the country, Sukarno, translates thus: 'Su' means good, and 'Karno' is derived from 'Karna', a character from the Indian epic, the *Mahabharata*. The name of his daughter, Megawati Sukarnoputri, (President from 2001–2004) translates as 'cloud goddess, the daughter of Sukarno'. Following her as President, for a decade from 2004, was Susilo Yudhoyono, a former army general whose name translates (appropriately!) as 'well-behaved knight/warrior'.

Names of Indian origin are not uncommon in other countries of the region too: for example, Thailand's deeply revered emperors are all called 'Rama' (the current emperor is Rama X), after the Hindu god of the same name. All over Southeast Asia, temples provide evidence of the centuries-old impact of Hinduism.

The influence of Buddhism, a religion that originated in India and has propagated its culture across many countries, spans the Asian continent from Japan to Central Asia. The colonial shipments of indentured labour have taken Indian culture even further afield: from Fiji in the Pacific Ocean, to as far as Trinidad and Guyana in South America. All this happened in the natural course of events and not as a result of any planned effort (in contrast to the British cultural push in various parts of the world, including India). As a result, till recently, India had rarely, if at all, leveraged the potential of these deep cultural links.

These extensive forays across land and sea would not have been possible without the 'technology' of boats or ships and of navigating uncharted oceans to faraway destinations. Even in those distant times, technology was the carrier of culture. Today, India's soft power in these countries owes much to those technology-enabled cultural voyages of bygone generations.

In the last half century, the Indian diaspora, or persons of Indian origin (PIOs), has grown rapidly across many countries. It now numbers an estimated 16 million, with the US, Saudi Arabia and the Persian Gulf region being the most popular host countries.[58] Economic betterment has been the major driver, but technology (cheaper, easier and high-frequency air connectivity, coupled with communication links) has also been a factor in making such migration less painful, by reducing the sense of being cut off. Now, migrants stay in frequent touch with their friends and relatives in India and, in conjunction with regular trips back to India, this considerably reduces the sense of isolation, loneliness or feeling uprooted. They, along with non-resident Indians (NRIs), carry with them the culture of India, which slowly diffuses into their adoptive community.

One example is cuisine. Indian cuisine was not always widely available. Till a couple of decades ago, getting Indian food in Europe—especially vegetarian—was not easy. I recall colleagues subsisting on little more than chocolate shakes and French fries from McDonald's! However, with a growing number of Indian immigrants in a given area, inevitably a restaurant sprang up to meet their needs, catering especially to the 'home food' that they missed. Over time, the local (indigenous) population began to try out this food and take a liking to it. As a result, restaurants serving cuisine from various parts of India have now become common in many countries abroad and, in most cases, have a large non-Indian and/or local clientele.

Over the years, Indian restaurants abroad have evolved, from functional, simple and often stark places serving only Indian food, to sophisticated, high-end and high-cost restaurants for fine dining. Indian cuisine is yet to achieve the global popularity of pizzas (Italian) and burgers (American), but in London, chicken tikka masala is probably easier to get than the traditional fish and chips! At Wimbledon, the strawberries in the traditional strawberries-and-cream are, I was told, from India. As a Britisher told me: 'At one time we had conquered India; now, you not only beat us at cricket, but you have conquered our stomachs too.'

Moving from cuisine to cars, the very British 'Jaguar' is now owned by an Indian company (Tata), which was the biggest industrial employer in the UK in 2014[59] and, even today, is amongst the largest. Clearly, Indian presence in the UK is not limited to the diaspora, but is also quite pervasive through investment, food and culture. While there is often criticism in India about the colonial mindset of the elite or the colonial hangover in our administration and legal systems, the reality is that there is also considerable Indian influence in many facets of British life. As former British Foreign Secretary, Boris Johnson, told a gathering at a celebration marking seventy years of Indian independence: 'We, in the UK, are the beneficiaries of reverse colonialism.'[60] This exemplifies India's growing soft power.

Another telling example of India's growing influence is in the US, where the Indian migrant—whether on a temporary work visa, or settled—has been a carrier of culture. Here, though, it is the values and work ethics of Indians, more than their cuisine, which has been widely noted and is greatly respected. Indians are seen as friendly, hardworking and well educated. Their success and prosperity have brought respect, with the occasional racist attacks on them being widely seen as undesirable aberrations. Yoga and Buddhism have moved from elite circles to the masses. Meanwhile, the perception of India as a country has also gone through a radical shift. When I first went to the US in the early '70s, India was yet widely viewed as a 'basket case', surviving on aid from abroad (the Aid-India Consortium, comprising of thirteen developed nations and the World Bank, gave aid to India in the '50s and '60s), with little hope of real economic growth. The Public Law 480 (the US public law for grain supplies abroad) shipments of food to India in the mid '60s were the genesis of the disparaging 'ship to lip' comments about India's dependence on imported grain. This resulted in the perception of India as a land of acute poverty, overpopulation, dirt and disease. The second most populous country in the world was so unimportant to the general American that it was nowhere on their radar. I remember being in the US (long before the days of online news) and scouring all their newspapers, hungering for even a tiny scrap of news about India. There was none at all—for weeks, at times—thereby signifying a complete lack of interest.

Today, media around the world have regular reports about India. In the US, India gets extensive coverage. More importantly, there has been a dramatic image makeover. Thanks to rapid economic growth and India's technological prowess— especially in atomic energy, space and IT—the country now enjoys a positive image in the public eye. This is exemplified by a personal anecdote (hackneyed and much used, but true and telling), based on experiences that others, too, have had. When one travelled abroad (a rarity in the '70s), the first real contact with a foreigner was at the immigration desk. For most Indians—and certainly for a twenty-something greenhorn like me—this was an intimidating experience. One would walk up to the immigration desk (after a tiring hour in the queue, following an eighteen- to twenty-hour journey), worried about whether all their papers were in order, and placed their passport on the counter. I still remember the look on the official's face when he saw it was an Indian passport. There was a scowl, and I could sense the disdain and suspicion, even as I could almost hear him say, 'Indian! Must be trying to sneak in… Better check his papers really carefully.' A grilling followed—

about my background, purpose of visit, place of stay, return plans, etc. The hostility and rudeness were palpable, and I really heaved a sigh of relief when the passport was stamped for entry. Now, for well over a decade, the story is very different. When I place my passport on the desk, I am greeted with a smile, and asked: 'Indian? Welcome, Sir. Are you from the IT industry?' One can sense the warmth and the genuineness of the greeting. Questions are few, and the passport is stamped with a parting wish: 'Enjoy your stay, Sir.'

It is not as if all the hostility and suspicion about fake papers have evaporated overnight. However, in an overwhelming majority of cases, there is friendliness and comfort. Even dislike is tinged with a measure of respect. This anecdote, in a way, reflects the metamorphosis in the image of an Indian—from that of a frail, indigent person with a begging bowl in hand, to that of one who's 'cool' and knowledgeable. This change is both the cause and the effect of India's soft power, with technology and technological capability playing no small role.

India's soft power has been growing in other countries too. Indian movies, especially the Hindi ones, have long been popular in a large number of countries. The term 'Bollywood' has increasingly become the shorthand for a typical Indian commercial film, especially with its plentiful song and dance sequences. It has become a widely used and recognizable moniker, indicative of the global visibility of India cinema. Raj Kapoor and his films may be only on the periphery of memories in India, but they are remembered even now in Russia.[61] Shah Rukh Khan's films are popular in Germany and he has a huge fan following there.[62] A few years back, I had great difficulty entering my hotel in Cologne, because of a massive crowd of screaming teenagers. I soon discovered that they were there only to get a glimpse of Khan. It is noteworthy that a majority of the crowd comprised ethnic Germans. Equally inexplicably, a few Tamil films have turned out to be great hits in Japan, and Rajinikanth has become very popular there. The cool-as-ice, villain-vanquishing superstar of many a Tamil film, with a massive fan following in Tamil Nadu—and elsewhere, too, in India—has reached cult status. Each new film of his is awaited more eagerly than even the first rains. This is not surprising in film-besotted India, but what is unexpected is that there are active Rajinikanth fan clubs in Japan too! A report says that one such fan club in Tokyo has about 3,000 members, and that there are similar clubs in Osaka and Kobe too.[63] Why Rajinikanth's loud and action-oriented films are popular in quiet, orderly and staid Japan is, indeed, a mystery. Possibly, it is this very contrast—between the Japanese way of life and the action-fantasy world of these films—that appeals to them. While his later films have not been as successful, Japanese fans continue to keep the clubs alive and imitate his trademark style and gestures; a few have even learnt Tamil. It is noteworthy that these films, especially in their action sequences, use a great deal of technology—for example, in animation or visual effects. Therefore, technology also aids their popularity.

Some Japanese have now gone beyond fan clubs and set up a South Indian food catering business, with vada-sambar being served in the traditional manner—on banana leaves. Also, dance schools have come up to teach Indian movie-style dancing. This twenty-first-century cultural impact of India builds on the age-old connection made through Buddhism. These constitute the elements of India's soft power and facilitate the geopolitical confluence that is emerging for strategic reasons.

If Japan, with its insular nature and absence of a substantial Indian diaspora, is an unexpected market for Indian films, another surprise is China. Traditionally, the Chinese have mostly looked down on other cultures, but even today, ancient Buddhist saints from India are revered in China. Now, it seems that India is again making inroads into the country through its films. This is definitely indicative of the potential of films as part of India's soft power.

China restricts foreign film releases to thirty-four per year, and most of this is taken up by Hollywood. Yet, over the last few years, Indian cinema has been making steady forays into this very large market. China's favourite Indian actor is Aamir Khan[64], who is known for his issue-driven movies, many of which have done phenomenally well in the country. Till 2018, the three highest-grossing Indian movies in China have been *Dangal* (₹13 billion[65]), *Secret Superstar* (₹8.1 billion[66]) and *Bajrangi Bhaijaan* (₹3.13 billion[67]). Khan has acted in, and produced, the first two. As is evident from its Chinese box office collection, *Dangal* took Indian cinema to an altogether new level in China. One result of these success has been a growing interest on both sides to see how more Indian movies can enter this big market (predicted to become the biggest in the world, overtaking the US by 2020[68]), and also exploring co-productions. Clearly, Indian movies have made significant inroads into the Chinese market, demonstrating India's growing soft power.

TV: A NOT-SO-SOFT SOFT POWER

Like films, TV, too, is a vital component of soft power (probably even more impactful than films, thanks to its technology-enhanced reach). Two examples highlight the importance and recognition of TV and communications technology as powerful tools, especially during confrontations between nations.

The first is from the war in the Balkans due to the disintegration of Yugoslavia. On 23 April 1999, during the Kosovo War, the North Atlantic Treaty Organization (NATO) forces bombed the headquarters of Radio Television of Serbia (RTS), severely damaging it and leading to the death of sixteen employees. RTS was known to have been broadcasting Serbian nationalist propaganda, which, many felt, demonized ethnic minorities and legitimized Serbian atrocities against them. NATO, too, justified the bombing on this basis, with the contention that RTS was making 'an important contribution to the propaganda war, which orchestrated the campaign against the population of Kosovo'.[69] On the other hand, organizations like Amnesty International and individuals like Noam Chomsky strongly criticized the attack.[70, 71] The incident highlighted the perceived importance of media in war, through its ability to influence people.

The second example is more recent and of Qatar. Following the fallout between Qatar and a group of countries led by Saudi Arabia, the latter imposed sanctions on the former in 2017.

Qatar was accused of siding with and instigating opponents of the governing regimes in the latter countries, and of supporting the Islamic State (IS) extremists as well as Iran. The group of countries then, put forth a set of conditions that had to be met for the sanctions to be lifted. A key condition was the shutting down of the Qatar-based broadcaster, Al Jazeera. The demand to shut down its broadcasts is a rare instance of the formal recognition of the key role played by media and is, in many ways, a left-handed compliment to the organization. The soft power of Qatar, through Al Jazeera and its apparent ability to influence people (especially in Egypt, Saudi Arabia and United Arab Emirates), seems to have riled its large and strong neighbours. Much of this power comes from its wide reach, made possible by the technology of satellite broadcasting.

A NEW 'WEAPON'

Even militarily strong countries acknowledge the importance of soft power and are increasingly focussing on this. One such country is China. Despite having the largest standing army in the world, it actively espouses and uses non-military means. Its 'three warfares' strategy comprises the use of media, psychological warfare and legal stratagems.[72] The Central Military Commission (CMC) approved the guiding concepts for its army, as far back as 2003, and have used the three warfares—with visible success so far—in the South China Sea dispute (though its military might undoubtedly provided the backing that amplified its psy-war and enabled it to brush aside the United Nations Convention on the Law of the Sea [UNCLOS] tribunal ruling against it). This strategy has also enabled China to win over a major opponent—the Philippines. In its confrontation with India over Doklam (which is at the India–Bhutan–China tri-junction), it seems to be following a similar strategy: integrating elements of the 'three warfares' with the unstated backup of its military might.

Communications technology is the essential and underlying means that enables and extends outreach. Broadcasting—first by conventional methods and now via satellites—has long been a powerful way of conveying information on a mass scale and seeking to influence people. In addition, technology has now made available a host of other means such as social media and other over-the-top (OTT) applications that can be received on computers or even on mobile handsets. The centuries-old print medium, too, now uses a great deal of technology in gathering, editing, printing and distributing information.

As countries become increasingly dependent on digital infrastructure (networks, computers, data and applications), any disruption in this could result in chaos and even disaster. Cybersecurity is, therefore, critical. In recognition of this, countries are developing both defensive and offensive capabilities in the cyber realm. Cybersecurity has, consequently, become an extremely important element of soft power and a key 'weapon' in a country's armoury. India, with its immense software capabilities, is well placed to become a leader in this new dimension of warfare.

Today, soft power is a necessary complement to, and an enhancer of, military capabilities, giving some an asymmetric advantage. Maximizing and leveraging soft power is the goal of every country, especially at a time when direct military warfare is generally frowned upon and often unlikely. The above examples indicate how technology is a multiplier—and sometimes an enabler—of soft power. This is important for India as its yet-underutilized soft power can be a real force only if it is amplified by technology (and backed up by military power).

Though few countries are replacing military formations with music bands, or firepower with films, a slow shift in emphasis is discernible. It is possible that in the battle for domination, weapons of mass communication may replace those of mass destruction, and the mind may win over munitions.

7

TECHNOLOGY OF THE FUTURE, AND THE FUTURE OF TECHNOLOGY

Opportunities and Dangers Ahead

Technology, as noted at the outset, has always influenced the course of human civilization. Overall, it has been a force for good, helping to reduce the rigour of human labour and bringing about major changes in the way humans work, interact and use their leisure time, thereby making life easier. It has been the biggest factor in raising productivity in all spheres and, thus, increasing incomes. Over the centuries, every now and again, it has created completely new products—for example, ships, trains, automobiles and aeroplanes, which have facilitated the movement of goods and people. The consequent boost to trading has been an important driver of global prosperity. The steam engine, electricity and electronics created revolutions on the farm and within the factory. Ever better connectivity—evolving from the telegraph, to the voice, to the video—made the world a 'global village'. Dramatic as it was, all this change in the life of humans, and the prosperity it brought, seems but a prelude to what we have witnessed in the last few decades, and the prognosis for the immediate future.

At the same time, before talking of a shining future, it is necessary to acknowledge the pain that technology has sometimes brought. The brains that conceived and realized so many wonderful technological products also created armaments. Inevitably, these weapons were used to fight wars and caused the death of millions of people. The minds that unravelled the mysteries of the atom, which led to major advances in science, were also the ones that devised the atomic bomb. Its use in Hiroshima and Nagasaki in Japan in 1945 led not only to a massive death toll, but also horrifying sicknesses amongst thousands of survivors. This has, so far, been the single most destructive use of technology. Sadly, nuclear weapons—which are even more powerful than those bombs—have been made; fortunately, none has yet been actually used in battle, though these weapons of mass destruction continue to proliferate across the world. Meanwhile, scientific research in various fields, including chemistry and biology, is being tapped to create new genres of armaments. Chemical weapons (CW), like phosgene gas, were used over a century ago, in the First World War, and new versions of both CW and biological weapons (BW) are known to have been developed and occasionally used even in recent years.[73, 74] Global treaties in the field of nuclear weapons as well as CW and BW aim to curb their use, even though these are often unequal as they seek to perpetuate the dominance of the powerful. While the idea of international treaties to limit the spread and use of such weapons is a worthy one, their real efficacy—especially in times of crises—is yet uncertain.

Meanwhile, many new technologies—GPS, data analytics, new materials, AI and robotics—are being used to multiply the effectiveness of weapons or of the individual infantryman. For example, GPS technologies are being used to identify and pinpoint the precise position/location of enemy targets. Unmanned aerial vehicles (UAVs) are another example of an integration of a number of these technologies. Beginning with aerial observation, UAVs now come in a wide range of sizes and have a variety of uses. Some are used for direct-offensive purposes, like dropping bombs or firing missiles (sophisticated technologies enable them to pinpoint a target, like a specific car or a room in a house, and guide missiles/bombs to them); others, such as drones, are miniaturized to the size of a bee and used for surveillance, sending back pictures and audio from areas where access is restricted.

Fortunately, many of these developments are being modified and adapted for less destructive purposes as well. UAVs are now extensively deployed for the following: monitoring remote and inaccessible areas (forests, for example); during floods, when access becomes a problem; monitoring crops to ensure a better harvest; locating and tracking the causes and extent of pollution; and mapping routes for power lines or pipelines. However, the fact is that technology in this field, as in many others, is being driven by the needs of the military or the security agencies.

Space technology is another example of an area where

military needs drove developments. Though the Outer Space Treaty bans the militarization of outer space and the Moon Treaty bars military bases on the Moon, anti-satellite weapons have not only been developed, but actually tested by some of the major powers[75], and could be used to knock out a large part of an enemy's surveillance and communication networks. Laser beams from space could be used to destroy key infrastructure on Earth. In addition to these directly destructive purposes, satellites can be used (and are already being used) as force multipliers for troops on the ground, providing communication, control, precise location information, weather reports and intelligence (like pictures of enemy formations or targets). It is in recognition of these immense capabilities that countries have been directing bigger proportions of their defence spending to space technology. In June 2018, Donald Trump announced the creation of a new 'Space Force'[76], an independent branch (distinct from the other five) of the US armed forces—a step beyond the already existing Space Command, which is part of the US Air Force.

However, there is also a silver lining to these developments as many of them are capable of dual use, both for military and peaceful purposes. Space provides a good example of how technologies developed for military purposes have helped civilian and peace-related causes. GPS was developed as a guidance system for missiles and continues to be used to that end. At the same time, the civilian use of this technology has now become extensive and commonplace. Satellites used first for military surveillance or spying are now used for a vast array of civilian purposes. These include crop-health monitoring, yield prediction, mapping land-use patterns, urban planning, locating optimum areas for fishing and measuring snow-melts to predict water availability.

In most countries, the civilian space programme was an offshoot of the military one. India is a rare, if not unique, exception. Here, the space programme and related technologies—including launch vehicles and satellites—were developed primarily for applications in areas like telecommunications, TV, weather forecasting, crop assessment, land-use planning, and disaster monitoring and mitigation.

BATTLEFIELDS OF THE FUTURE

Recent years have seen the growth of so-called 'asymmetrical warfare', with small or lesser-armed forces able to take on a larger and more powerful enemy. Guerrilla warfare has long been one example of this, where small and highly motivated groups can take on a big enemy through the stratagem of well-planned surprise attacks. As mentioned earlier, in Vietnam, the Việt Cộng was able to inflict significant damage to the hugely more powerful US military force. The Soviets faced a similar problem—and subsequent defeat—in Afghanistan in the late '80s, leading to a final and complete withdrawal of Soviet combatant forces in 1988–9. Such guerrilla warriors have always depended greatly on support from the local civilian population, which allows them to not only move around undetected, but also get sensitive information from additional eyes and ears.

Now, there is a new form of such asymmetrical warfare that

does not need local support—cyberwarfare. A large part of this is through cyberattacks, with motives ranging from harassment (by defacing a website or making it inaccessible) to theft (by hacking into an account and transferring money). As mentioned earlier, sometimes the attack is aimed at extracting payment (ransom), with the transaction done in cryptocurrencies (most of which are difficult to trace). Though countries and organizations are continuously shoring up their defences against such attacks, the fact that more than a few continue to breach these defences is indicative of the growing sophistication of the attackers. Most of these attacks emanate from one—or more commonly, a small group of—accomplished computer expert(s). They don't need any local support, unlike guerrilla warriors, and depend only on their own skills, some computer hardware and electronic connectivity. With these, a handful of expert hackers can successfully attack a very large organization—even a government.

There are a few groups that go beyond defacing, attacking and demanding ransom. They seek to disrupt critical infrastructure in sectors like finance, health, transportation and utilities (water and power). The fact that, in an ever increasing number of countries, infrastructure is now heavily dependent on IT makes it particularly vulnerable to cyberattacks.

There are ways in which such attacks can be mounted so as to cause not merely disruption, but also destruction. Thus, it is not just a matter of, say, shutting down the gas supply pipeline through a cyberattack, by turning off supply valves; there are ways by which an attack can possibly use the gas to cause an explosion. One is unlikely to read about such an event because, even if it has happened, countries do not want to let it be reported. This is, of course, ideal fodder for movies. For example, in the box office hit *Die Hard 4.0*, the antagonists create chaos, engineer a gas explosion and more, by hacking into computer systems using skilled cyberwarriors. The protagonists are able to counter them only by tapping into another hacker's skills.

The primary requirement for mounting such attacks is skill of a high order. The equipment required is minimal and no other material is necessary. This makes it entirely possible for any small group of computer software experts, with exceptional capabilities in the right areas, to even take on the might of a State. Such non-State players can—at least in theory—not only paralyse a country, but can do this cleverly enough to be almost untraceable.

There are groups—not necessarily backed by a country—that are creating mayhem through chemical, biological or conventional weapons. However, the destruction that they can cause is generally geographically limited. Also, it is comparatively easier to trace them, since they are dependent on a supply chain—of weapons, or chemical and biological agents. In the less likely case of using radioactive weapons, traceability is even easier, through equipment that can detect radioactivity. Further, there is also a money trail, since expensive purchases have to be made. Cyberattackers are able to overcome almost all these constraints: no raw material or physical goods (beyond off-the-shelf computer hardware) are required; no major financing is involved; there is no need of any supply chain; and damage can be geographically widespread (for example, disrupting the banking or financial infrastructure of an entire country). In fact, independent non-State cyberattackers who disrupt or destroy infrastructure are now a reality.[77]

There are now growing suspicions, though, that some of the cyberattackers are also State-supported.[78] This is not a surprise, because many recognize that future wars will depend upon cyber capabilities. Even in times of 'peace', one can use these capabilities to test cyber 'weapons', and the opponent's defensive competence, through attacks on a non-attributable basis, since tracking the source of attack is difficult and linking it to formal State machinery even more so. One example is the famous Stuxnet, a software bug that apparently disabled a large number of Iran's centrifuges in 2010.[79] These were being operated to enrich uranium to bomb-grade levels. Rumours abound about this being done by Israel or the US, but there is no concrete proof. While Stuxnet is probably the most high-profile incident of its kind, stories do go around about a number of other attacks that have crippled systems of countries for extended periods of time. These include the attacks on Estonia in 2007 and Georgia in 2008.[80, 81] Fingers were pointed towards Russia but, again, no proof or trace was found.

In other areas of warfare and attacks, much is achieved through credible deterrence, with the attacker always in danger of a counter-attack. However, in the case of cyberattacks, a strategy of deterrence is difficult because attacks are generally neither attributable nor traceable. In some cases, the attackers could well be disgruntled groups, genuinely independent of any State. This, of course, also gives States, so inclined, an easy cover to hide behind.

WHY TECH TRIGGERS TREPIDATION

Despite the above, and its many negative uses for war and destruction, technology has historically been a positive factor driving the development of civilization. However, many now look at the technological future with trepidation. The accelerating pace of change and some of the emerging technologies are creating deep concern. Many fear that robotics and automation may lead to a loss of jobs. It is partly for this reason that some countries have begun to seriously consider a Universal Basic Income (UBI)[82]—a guaranteed income for each citizen, for life, which will ensure a kind of social safety net for all. This is not a concept merely for rich countries whose economies may be able to afford this (some have even begun pilots, with a limited number of people being provided a UBI); even in developing countries like India, this idea is being seriously discussed. In fact, the concept was advocated in the government's Economic Survey, 2016–17.[83]

If and when there is a UBI on a larger scale—whether triggered by a loss of jobs or as a social security measure—we will face another question: with an assured income but no work, how will people spend their time? Most people enjoy leisure and use it for a variety of purposes—from travel, to hobbies, to spending time with family and friends. This, however, is usually a break—an interregnum from work. A life of permanent leisure is a very different thing. Even the thought of 'doing nothing', on a continuous basis, may seem depressing to many people. Could it, then, affect people's mental health? Will they be tempted to substitute the excitement, thrills and failures of work life, with drugs or crime? Or will they spend their time painting and writing poetry? Creativity may well flower—or suicides may increase.

New technologies have also given rise to other concerns. One is a widening inequity, as the rich get smarter (thanks to bio-implants) and so become even richer. Another, more grave worry is about AI taking over and making humans second-class citizens. Many feel that some human capabilities are unique, including creativity, innovation, disruptive thinking and drawing analogies between various disciplines. Could a machine have deduced the laws of gravity on the basis of seeing an apple fall, or wondered why the sky is blue and deduced the Raman Effect from that? Who knows, maybe at some point of time, machines will be capable of such leaps of thought? What does seem certain is that in the not-too-distant future, machines will be capable of doing much of what most people do now. Meanwhile, humans may evolve and develop other special capabilities, creating altogether new kinds of jobs.

Is such optimism merely whistling in the dark? Probably not! After all, through all the past technological revolutions, including the advent of the electronic/computer age, humans have continued to stay on top and also create so many new jobs that we are, by most accounts, witnessing the lowest unemployment levels in recent history. It is true that the pace of technology development is almost exponential and that the combination of progress in different fields (especially robotics, machine learning and AI, biology and genetics, 3D printing and new materials) is leading to synergies that could cause radical changes. We are probably at the cusp of dramatic and disruptive developments; yet, history dictates optimism.

SEEKING REFUGE IN NEW HABITATS

Amongst the many changes being talked of, the most radical is the concept of creating extraterrestrial settlements. Studies have looked at the details of setting up permanent human habitations on the Moon[84] and on Mars[85]; some have even speculated about such settlements on suitable planets outside our solar system. This is, of course, the basis of many a science fiction story. However, it seems that this may well become a reality within this century itself and people have already begun identifying specific locations. One probable site for future human habitation is a giant volcanic 1.7-kilometre-long cave beneath the Moon's surface, discovered by researchers at ISRO's Space Applications Centre in 2010, through an analysis of the 3D images from the Chandrayaan-1 spacecraft.[86] This cave is a hollow tube, created by ancient volcanic lava flows, with a cavernous mouth—about 120-metres high and 360-metres wide—and a roof estimated to be 40-metres thick. This may provide lunar explorers a natural shelter, protecting them as well as their instruments from radiation storms and extreme variations in temperatures on the Moon's surface.

Though the idea of living on another planet (or on the Moon) may seem revolutionary, is this very different from what Europeans may have felt when they created their first settlements in India, Indonesia or South America? Ships enabling transoceanic voyages were the technology that drove many of the earlier settlements; spaceships enabling trans-planetary voyages are the technology that may help create new human settlements. Conceptually speaking, how different is a human colony on the Moon from the first permanently manned stations in Antarctica? All such settlements share a few distinctive common features: remoteness and isolation from the point of one's origin; being mostly cut off from similar kinds of people, except those in the settlement itself; a real (or perceived) hostile environment outside the settlement, where stepping out unprepared (whether in Antarctica or on the Moon) may lead to death; and long-delayed and often tenuous communication links with one's 'home'. In that case, is the thought of an extraterrestrial settlement all that unique or new to humankind?

There are many reasons why humans may want to look at new settlements even in strange, forbidding places. One, of course, is a desire for exploring the unknown—something that has been with us from the very beginning and is probably embedded in our DNA. When asked why he chose to climb Mount Everest, George Mallory, the famous mountaineer of the '20s, responded, 'Because it's there.'[87] This typifies a basic trait of humans: to undertake arduous and dangerous missions purely out of an innate sense of adventure and exploration.

There are more pragmatic and compelling reasons, too, for settlements beyond Earth. These include doomsday scenarios of a large-scale nuclear war that not only causes death and devastation on an unprecedented scale, but results in enough radioactive contamination to make large parts of the planet uninhabitable.

Beginning to move away from the planet may, therefore, be a necessary escape.

Another scenario is a danger from space: a large asteroid crashing into the Earth. A report claims that the extinction of dinosaurs occurred due to one such collision 66 million years ago, when a 7.5-mile-wide asteroid crashed into the ocean off the coast of Mexico.[88] It is not inconceivable that another major event of a similar nature may make the Earth uninhabitable for humans.

As for many other problems, technology offers possible solutions for this too. One is to 'attack' the asteroid with missiles and bombs before it reaches the Earth, so as to either change its path (hence avoiding a cataclysmic collision) or break it into smaller fragments that will burn up in the Earth's atmosphere. Scientists are already tracking big asteroids to see if any are headed directly towards us, so as to have advance warning and take required action. Whether or not such defensive action is practicable is yet to be seen.

Another serious concern is the danger of climate change. The average global temperature has already seen a steady increase, which is indicative of global warming. A major cause is pollution resulting from growing industrial activity, especially greenhouse gases like carbon dioxide. A study indicates an increase of over $2°C$ by the year 2100.[89] In fact, the report published by the Intergovernmental Panel on Climate Change (IPCC) in 1990 had predicted an increase 'during the next century (by) about $3°C$ per decade (with an uncertainty range of $02°C$ to $05°C$ per decade).'[90] If the rise in temperature is not controlled, the consequent impact could be catastrophic: melting of substantial parts of glaciers as well as polar ice caps; rise in sea level, which would inundate coastal cities and areas, and possibly obliterate small islands like the Maldives; huge loss of crops; droughts and floods; and a serious impact on health due to diseases caused by a combination of these factors. India, according to a recent report, could be the country worst hit by the falling crop quality the world over, due to rising carbon dioxide levels.[91] High atmospheric levels of carbon dioxide caused by human activity lead to less nutritious crops, with lower concentrations of protein, iron and zinc. This would result in an estimated 50 million more Indians facing zinc, iron and protein deficiency. This is particularly worrisome because, as the report says, even now, as many as 35 per cent of Indian children are underweight (compared with a developing country average of 20 per cent), 38.4 per cent are stunted (low height for age) and 21 per cent wasted (low weight for height).

It was these concerns about the effects of pollution that drove nations to sign pacts like the Kyoto Protocol (1997) to limit the emission of greenhouse gases, and the more recent Paris Agreement (2015). The efficacy of both is limited by the fact that the biggest polluter, the US, is a non-participant.[92] It hadn't acceded to the Kyoto agreement and also withdrew (in 2017) from the Paris accord, after having signed it.

Many people feel that climate change is already on an irreversible path and that it may be best for humans to plan a large-scale migration to habitations elsewhere. Alternatively, extraterrestrial settlements could be the base for all polluting industries, leaving the Earth a 'cleaner' place, which can slowly heal from the damages of the past. This scenario sees Earth almost like a 'dormitory' habitat, with much of the industrial activities being carried out elsewhere.

Thus, the largely technology-induced problems (war, climate change) and even the 'natural' one (asteroid strike) could be solved with the help of technology—by creating new habitats outside the Earth, and transporting humans to them.

INTERNAL CHALLENGES

As humankind looks at the possibility of radically new external environments, the 'internal' environment of humans may also see a big change. With organ replacement, artificial implants, gene editing and other such innovations, the man or woman of tomorrow may be, in many ways, almost a new species. Such bioengineering could create humans with exceptional physical and/or mental abilities: supermen and superwomen (though they will not be able to fly like their comic book/movie counterparts). With organ repair and replacement, and other developments in biology and genetics, humans could even become immortal. This raises many questions, both practical and ethical. Would one want a memory that can clearly recall everything? A long and healthy life may be welcome to almost, but would a never-ending one be as desirable? If humans do become immortal, will the issue about an overpopulated Earth become a more grave concern than it is today? Will the answer be to stop having babies?

Everyone may not be able to afford the costs of implants and the like. Those who can (the rich), will become physically more powerful and mentally more capable than others. This could give rise to vast inequities between those who can buy superpower capabilities and those who cannot. Further, the rich will be able to pay for the cost of creating 'perfect' designer babies (made possible by biogenetics), maybe leading to a new caste system, since it will begin at birth itself.

New developments and breakthroughs in the field of biology, genetics and bioengineering are very exciting; yet, at the same time, they are a cause for concern. Whether or not humans migrate to other planets, these internal issues will remain. If there are to be interplanetary migrants, it is logical that they will be the ones engineered to have greater physical and mental capabilities, to take on the rigours of creating extraterrestrial colonies in an unknown environment. For this reason, as well as the long journey (decades, or maybe even centuries), they may need to be near immortal. Will this migration of superior humans result in Earth becoming a colony of the outposts in the course of time, rather than the other way around?

ALIENS AHOY!

For some time now, scientists have been engaged in the search for extraterrestrial intelligence (SETI). There are installations that look into space and use instruments that constantly monitor radio frequencies most likely to be used by intelligent alien beings. So far, no signal has been detected. One explanation is that we may be unique in the whole universe. On a purely statistical basis, this seems unlikely, especially as recent discoveries indicate that there are more than a few exoplanets that have conditions similar to Earth, and so are capable of sustaining life as we know it. Similarly, with the presence of more than 2,000 confirmed planets, the possibility of all other civilizations being so underdeveloped that they have not yet discovered the use of radio waves for communication is statistically contradictory. This 'radio silence' from the rest of the universe is, therefore, difficult to understand. Is it that all other civilizations have destroyed each other? Is the course of life such that it self-destructs when it reaches a certain stage? Have all other civilizations developed non-radio forms of communication that we cannot yet detect? Or, is creation a non-statistical phenomenon and we are, in fact, unique? Technology makes search possible and increasingly sophisticated, but has no answers yet to these questions.

Some people also fear the detection of intelligent life elsewhere. While such a discovery would be really exciting, they feel it can pose a danger to—and possibly lead to the destruction of—humankind. This is a rather pessimistic view of the evolution of life and the sensitivities that develop along with technological capabilities. Yet, it can be argued that our own technological progress—from the earliest tools, to the sophisticated technologies of today—has not necessarily made us any less destructive. After all, despite the wonders that make life easier, healthier and wealthier, we have also moved from simple projectiles to sophisticated bombs and missiles; from being able to hurt only one person at a time, to killing hundreds of thousands with a single weapon. Will some advanced civilization find us, and squash us like pests or enslave us like the colonizers of the past? Or will they, too, be thrilled to find another civilization and share with us the wonders of their progress?

These are questions with no definitive answers, but they certainly provide much food for thought. They are profound, civilizational questions that look at the future of humankind as a species, and its survival on Earth or escape to habitations that they'll create elsewhere. Yet, there are also more immediate issues that technology throws up, which are not related to our exodus or the discovery of extraterrestrial intelligence (or their discovery of us). Excitement regarding new vistas, or fear regarding the same; a destructive weapon of war, or a tool for development; a means to foster a sense of community, or an echo chamber for bigots; a way to bring about good health for all, or the development of special, extraordinary powers for a few—technology is capable of providing all these. However, as the preamble of the Constitution of the United Nations Educational, Scientific and Cultural Organization (UNESCO) says, '…since wars begin in the minds of men, it is in the minds of men that the defences of peace must be constructed.' As experience indicates, the power of technology can be used both for good and for evil; to build and to destroy.

The choice is ours.

8

INDIA'S TECH TRIUMPHS

Translating Past Lessons into a Future Strategy

arlier chapters have discussed technology's impact on
areas that encompass a great deal of human activity
and also some of the larger issues that arise from the
progress of technology. Whether countries will continue to exist
as separate nations in the future may be doubtful, but till they
do, each one must see how it can best use the opportunities
thrown up by the emerging technological scenario. It is in this
context that India needs to put together a policy framework
and a strategy that will enable it to leverage its strengths and
build new capabilities to capitalize on emerging technological
developments.

An American scholar, who had studied Indian culture and
also spent much time here, once told me: 'You Indians are proud
of your heritage—and very rightly so—but you never seem to
draw on it. Most of your ancient monuments, instead of being big

tourist draws, are in disrepair and are vandalized. Past wisdom—encompassed in art, literature, parables and mythology—is not used as a source of learning for today's context.' This struck a chord, and I feel it may be worth drawing lessons—at least from our immediate past. Therefore, I thought that in mapping out a strategy and formulating policies, it would be instructive to see what can be learned from areas where the country has achieved a measure of success in recent times. In the sphere of technology, there are three areas in which India's success has been globally acknowledged—nuclear, space and IT.

SALUTING PIONEERING EFFORTS

In the area of nuclear technology, thanks to the foresight and leadership of Dr Homi Bhabha, a renowned nuclear physicist, India's initiatives began as early as the '40s. Recognizing the potential of this area, Dr Bhabha wrote to the Sir Dorabji Tata Trust in 1944, seeking funds for an institute to develop a research base. This took the form of the Tata Institute for Fundamental Research (TIFR). Set up in 1945, TIFR first operated from the campus of the Indian Institute of Science (IISc) in Bengaluru (then Bangalore) and moved to Bombay later in the year. The picturesque new campus near the sea, in Colaba, was inaugurated by then Prime Minister Jawaharlal Nehru in 1962. TIFR succeeded in its mission of attracting and creating a nucleus of researchers and developing human resources for the future.

After the creation of TIFR and with the possibility of going quickly beyond research, Dr Bhabha saw the need for a larger scale of work, with more focused goals. To do this, the Atomic Energy Establishment, Trombay (AEET) was set up in 1954 on the outskirts of Bombay, with enough land for substantial expansion. AEET was later renamed Bhabha Atomic Research Centre (BARC), acknowledging the role and contribution of Dr Bhabha as the father of India's nuclear programme.

As the programme grew and the strategic dimensions became clear, it was obvious that the government would have to take charge. Accordingly, within the government, a formal structure was created in the form of the Department of Atomic Energy (DAE) in 1954, and the Atomic Energy Commission (AEC) in 1958, with a common head reporting directly to the Prime Minister. However, sadly, Dr Bhabha, who had been selected to head both entities, died in the prime of his life, in an air crash near Mount Blanc in Switzerland in January 1966.

Research was complemented by design, development and engineering, including the creation of pilot plants at BARC. Where necessary (and possible), foreign collaboration was sought. While research and development (R&D) were key components, the broader goal of the effort was clearly directed at realizing practical gains—in this case, mainly through power generation. In keeping with this, the first commercial nuclear power station was operationalized in 1969. The development of weapons (nuclear bombs) came later, with a test (announced as 'a peaceful nuclear explosion') first in 1974, and then—a more definite step, with multiple explosions—in 1998. Meanwhile, other applications of nuclear technology were also developed— from sterilization of medical supplies and of potatoes and onions (to extend their shelf life), to the treatment of cancer. Thus, the core of the nuclear programme, like the space programme later, was driven by non-military (civilian and developmental) needs.

When I joined the AEC in 1968, the atmosphere was charged. Of course, it was my first job and at twenty-one, I was easily impressionable. Yet, there truly was excitement in the air. India's first power reactor was being set up in Tarapur, north of Mumbai, with US collaboration (it was commissioned in October 1969, and is said to be the world's oldest nuclear power plant yet in commercial operation). We were conducting studies on large-scale nuclear-industrial and agro-industrial complexes, which would co-locate large nuclear power plants and high power-consuming industries (like those that manufactured aluminium), or use the power to pump water for agriculture. New nuclear power projects and heavy water plants were being planned; studies on optimizing the power mix (between coal-thermal, hydroelectric and nuclear) were under way; the corporatization of uranium mining and production was taking place; and 'professional' management was being introduced. In addition, a decade profile (a ten year plan for the '70s) was being prepared. To be in the midst of all this and to see it from the vantage point of the office of the Chairman (as his Special Assistant) was exhilarating. The icing on the cake was the opportunity to work very closely with Dr Vikram Sarabhai, who had succeeded Dr Bhabha.

I worked with Dr Sarabhai for just about three and a half years, before his untimely death in December 1971, and had already begun to focus more on the space programme than the area of nuclear technology. Yet, the unique advantage of working closely with the top boss was the insight one got into various approaches, policies and strategies, and the thought processes that influenced them. On the basis of this—and, of course, with the advantage of hindsight—it is possible to identify the key elements that led to the success of the nuclear programme. A vital feature was the human resource base: attracting, developing and motivating the talent that the programme required. A long-term plan, executed through an appropriate organizational structure and managerial processes, was also crucial. Specific elements that made the programme successful, and which could be used to lay down a strategy for the future, include:

1. Identifying a technological opportunity that was appropriate and important for the country.

2. Developing the human resource base required for this technology. To do this, and to kick-start the use/ development of the technology, institutional structure, processes and facilities need to be set up, which will attract, retain and motivate talent from India and abroad. Selecting the right people and building institutions around them is also important.

3. Creating interdisciplinary teams to evolve a vision and develop a long-term plan, as well as detailing the concrete steps necessary to achieve it, including specific deadlines and targets.

4. Initiating collaboration with other institutions and countries, so as to leverage and use their capabilities to further India's goals in the field; importing technology and

equipment (as necessary), while developing indigenous capability; and looking specifically for know-how, know-why and training.

5. Ensuring that sufficient resources (particularly, funding) are directed towards this area and that it has a high-level champion—preferably within the government.

6. Ensuring that the leadership—certainly at the top, but even throughout the hierarchy—is comprised of scientists, technologists and/or professional managers with long-term interest or commitment in the area, and not transient administrators.

These are not necessarily in order of importance, and some actually took place simultaneously. However, though not comprehensive, collectively they do indicate the broad strategy that influenced the development of the nuclear effort.

The story of India's space programme bears many a striking resemblance to the nuclear one. I was, indeed, fortunate to be involved with this, too, from a very early stage. At that time (and until 1972), space activities were overseen by the DAE, and Dr Sarabhai, as Chairman of AEC, was concurrently in charge of the field of space. When he first told me (in 1969) about opportunities to work in the then relatively new, amorphous and nascent programme, I was excited but also somewhat reluctant. My hesitation stemmed from the fact that I would have to move from Mumbai, a bubbling-with-life city that was ideal for young, single people, to Ahmedabad—perceived then as a non-cosmopolitan, somnolent place. Of course, I had spent two very enjoyable years studying there, but that was because

I lived in an ivory tower—my only contact with the city was through restaurants, movie theatres and the railway station! However, I did begin my association with the field of space in 1969, and found it interesting enough to overcome all my doubts and move to Ahmedabad in 1970. My career there began with studies on, and planning for, the INSAT system. As a pilot for this, a year-long test was planned, using a National Aeronautics and Space Administration (NASA) satellite. This project—the SITE—has been widely acclaimed as a seminal one. In India, it took TV to remote parts of Odisha, Bihar and four other states, before it reached many major cities in India. INSAT and SITE are mentioned here because they exemplify the elements of strategy that drove the overall space programme.

Through the '70s, '80s and '90s, the space programme continued to make major technological breakthroughs and even operationalized new applications of the technology at an ever increasing pace. Soon after Dr Sarabhai's untimely death, Professor Satish Dhawan took charge of the space activities and the headquarters moved from Ahmedabad to Bangalore. He not only ensured continuity (of the vision and goals), but also the implementation of changes necessitated by the advanced stage of the programme. Subsequent chairmen of ISRO have each brought in their unique capabilities and ideas, while maintaining the overall vision and culture of the organization.

This century has seen consolidation, incremental capabilities, operationalization, and an expansion in the range and use of both technology and applications. It has also seen Indian probes go to the Moon[93] and orbit Mars[94]. All through, the programme has stayed focused on space applications and exploration, clearly concentrating on civilian—as opposed to military—needs. It has won global recognition not only for its achievements, but also for doing so in a frugal manner, thanks to innovations and a cost consciousness generally absent in government programmes.[95] The strategy was based on a clear vision, executed through structures that ensured autonomy and were based on a great deal of delegation. System thinking and planning, and multidisciplinary teams that included social scientists, were vital features, as were innovation, time and cost consciousness.

The key elements of the overall strategy can be summed up as follows:

1. Clearly articulated goals (mission), with targets and dates, based on a broad vision. This was communicated throughout the organization to create goal congruence and motivation. A long-term perspective (ten year plans) and an even longer vision were central to this.

2. Authority and responsibility were delegated; the integration of efforts went hand in hand with autonomy.

3. A culture was created in which milestones, targets and budgets (cost control) were important. Teamwork and frank peer reviews were an integral part of every project.

4. Administration and finance were support/service functions. Control and final decision-making were the prerogative of the technical/managerial leader of each project.

5. Failures were unfortunate but always acknowledged and treated as learning experiences. A detailed review

followed each failure, to determine the root cause and take corrective action.

6. The leader took the responsibility for failure, while success was always attributed to the entire team, including the project head. This was outstandingly and famously exemplified by Professor Dhawan, who publicly took responsibility for the failure of India's first satellite launch vehicle. When the next launch was successful, he stepped aside and gave full credit to the project director.

7. Partnerships and collaboration with other countries were entered into, right from the inception of the space programme (experiments with foreign 'sounding rockets' were India's first efforts in space), even as India's own capabilities were built.

8. Social scientists were a part of the organization and helped to ensure that applications were geared to the needs of people in rural India.

9. Maximum possible use was made of other organizations and industries for development and manufacturing, by transferring technology where required. The goal was not so much of meeting the immediate need, as about the vision of developing a space technology manufacturing industry in India.

10. System thinking (or end-to-end planning) was vital, and a key element for translating a broad vision into specific actions. Thus, the vision for satellite communications began by studying the design of the overall system, followed by a pilot (SITE) in which some elements

(ground hardware) were developed and made locally. The next step was an indigenously made satellite, which was finally followed by an Indian-made launch vehicle (designed for the needs of launching geosynchronous communication satellites).

As one can see, there are many common elements in the strategy of the nuclear and space programmes. An overarching element of the Bhabha–Sarabhai strategy was that applications paced technology; by and large, technological development was defined and driven by the already-planned application. Pacing requires that the application be ahead of technology development. Thus, nuclear technology development was focused on and speeded up by an application: a power reactor (the Tarapur Atomic Power Station) imported from the US. Similarly, the development of space technology was paced by an application: the INSAT system. It pulled along the development of satellites and launch vehicles, besides various ground segment elements. Such a process ensured that the technologies being developed are of immediate relevance, and aimed at meeting pre-defined needs.

The other common feature was a visionary and charismatic leader, who led the programme from the beginning, and one who had a strong backing from the top (in both cases, from the Prime Minister). Also, it is more than incidental that both organizations have their headquarters outside Delhi—the only departments of the Central government that are not based in the capital. This certainly helped to reduce the temptation and possibility of political interference.

THE DOT-COM BOOM

The third example of technological success for India is in the field of IT—specifically, software and IT-enabled services. While the two earlier examples (nuclear and space technology) shared many commonalities, IT is different in numerous ways. The most striking factor is that whereas the first two were government-led, the private sector played a major role in IT.

The '60s were the early years of the IT industry in India, and it was small and almost insignificant. In those years, software was a small, and generally free, adjunct to the large and sophisticated computer hardware that constituted the bulk of IT. The pioneering work in this field was initiated by Dr F.C. Kohli, a legendary father figure for the IT industry, and even though it was done in the prominent Tata group, it was but a minor activity. Today, of course, the flagship company of the group, Tata Consultancy Services (TCS), is recognized around the world. It accounts, by far, for the largest proportion of the group's profits and market capitalization.

Through the '70s and '80s, the software industry moved ahead slowly, building its capability and acquiring a limited number of foreign clients. It was as focused, if not more, on the small domestic market as the big global one. A substantial part of the expertise came from the engineers in the electronics industry, much of which was owned by the government. At the entry level, engineers—with a background in computers and mathematics, and trained in logical thinking—were found to be the most appropriate recruits. In the '90s, growth accelerated, particularly as customers abroad realized that the very large and appropriate talent pool in India could be tapped, and at a far lower cost than elsewhere.

This brought huge business to the Indian IT companies during the Year 2000 (Y2K) problem[*].[96] Many thought that India's IT boom would end with the closure of the Y2K issue, but this was proven wrong—as was the fear that the end of the so-called 'dot-com boom' would spell doom for the industry.

Riding high on the track record of making success of these two opportunities, Indian companies created a global market for themselves by leveraging lower costs as well as the country's human resource. In addition, a drive to meet the highest global quality standards was eminently successful, with many companies being certified at the highest level. This firmly established India's unbeatable triumvirate: the magic mantra of cost, quality and scale (of human resources).

Earlier, India had pioneered an innovative business model—one that combined outsourcing and offshoring. This new business model gained traction and momentum over the earlier 'manpower supplementation' one, in which Indian companies were basically recruiting, training and shipping human resources to companies abroad. Importantly, it enabled Indian companies to develop and cumulatively build expertise in various industry segments. For example, outsourcing (to India) by banks enabled the Indian IT companies to develop expertise in the banking sector, thereby facilitating the acquisition of new customers

---

[*]This concerned a coding issue that cropped up while transitioning all software and systems at the turn of the millennium—from 1999 to 2000.

from that sector, but also doing the next level of tasks in that area. Thus, they inched their way up the value chain, with a few companies even developing packaged software 'products' in the banking and financial sector.

The most successful and visible instance of value enhancement is in the segment that began as call centres and business process outsourcing (BPO). Given the abundant availability of educated and English-speaking people in India, a few pioneering companies set up their own facilities in the country to do some of their 'back end' work. This ranged from the simplest (data entry) to the more complex (calculating/ tallying figures for accounting or other purposes). More (usually fledgling) Indian IT companies and entrepreneurs saw an opportunity, and grabbed it immediately. In a few years, the BPO industry was booming, with annual growth rates in the twenty-plus percentage range. Work done in India quickly moved up the value chain, as customers gained confidence in the ability of Indian talent to take on and successfully deliver demanding and sophisticated work. Today, many multinational corporations (MNCs) do the most complex tasks related to accounts, finance, procurement and human resources, in India, and have large shared service centres here, catering to their global operations. More recently, data analytics has become an added expertise, used for market research, customer insights and even fraud detection. With the growing capabilities of, and confidence in, India, many companies have moved a whole range of functions (and not just tasks)—like accounts receivable, procurement and human resource management—to India.

Call centres, too, have seen a similar evolution in terms of work and value addition. They began with handling the simplest of queries ('What is my bank balance?' or 'What time is your flight to Paris?'), which only required the person at the call centre to look up the specific information on a computer and convey it to the caller. The skill required was rudimentary: reasonable computer literacy and some proficiency in English. To ensure an understanding of different accents and to be able to speak with a certain accent, companies organized special training for their call centre employees. As call centres evolved to the next level, some employees had to even undergo high-level training to be able to market a product to a caller or handle a complaint from them (the caller being a customer of the foreign company). There was a phase when companies wanted to make it seem like the responder (at the call centre) was from the home country (generally the US). Hence, some Indian call centres not only trained their staff to speak with an American accent, but also made them familiar with American culture. I visited a call centre that was working for a Chicago-based company, where the employees not only spoke with an American (mid-West) accent, but also responded to calls as Harry, Robert, Jane or other such common American names. Moreover, the walls of their office had posters or photos from the US and they knew not only the weather in Chicago that week, but also the football or baseball score of the latest games in the US! This (as an opening gambit, or while the required information was being pulled up on the computer screen) enabled them to establish a comfortable rapport with the caller and was intended to put the caller at ease. Fortunately, this pretence phase ended soon, with the easing of the so-called 'outsourcing backlash' that had hit a strident peak during the US presidential election campaign in 2004. Later, the training focused on speaking with a 'neutral' accent and the actual location of the call centre was not made a secret.

Call centres soon evolved to handle far more sophisticated tasks. Some companies, taking advantage of more than computer literacy and English, set up their help desk in India. This required the responder to often have advanced knowledge about the products or services of the company, necessitating expertise in computers or particular software packages at a level deeper than the caller (who might well be an expert in the area). Thus, highly qualified and experienced people worked in these call centres (and, quite naturally, resented the tag, since call centres had begun to mean low-end, low-skill places for people with no technical expertise). An evolution on these lines across the whole range of BPO activities prompted the change in nomenclature from BPO to BPM—business process management.

At the same time, innovations have seen the revenue model change—from time and material, to fixed cost, to a per-transaction fee, and finally to a risk/reward sharing, depending upon the type of work and the closeness of the relationship (ranging from that of a vendor to a strategic partner). Many Indian companies have been able to build excellent relationships with their customers by not only delivering to cost, quality and time specifications, but also by adding greater value. As a result, they have been able to mine deeper, and often spread wider, into the customer's business.

The evolutionary growth of the bread-and-butter services continued to bring success to Indian IT companies, and India remained the dominant global destination for outsourcing. There were—and are—new competitors: the Philippines, in particular, but also Vietnam and countries in Central/Eastern Europe. Indian companies have met this challenge in two ways: first, by working continuously on their competitiveness in delivering from India; second, by setting up call centres abroad, so as to be closer to the customer (in the US, the UK, Europe and Japan; and a few even in Central/South America and China) and/or to leverage any special competitive factors. For example, an affinity with American culture and the knowledge of American accounting principles gave the Philippines a special advantage. Likewise, high-end mathematical skills made Eastern Europe a favourable base for sophisticated financial and other analyses. In both cases, even the cost is comparable to India.

FUEL FOR TAKEOFF

Popular folklore has it that the IT industry grew and prospered because it was not on the government's radar. As the Minister for IT in 1999, the late Pramod Mahajan, often said (only half in jest), 'India has done well in two fields because the government has stayed out of them: IT and beauty' (the latter was in the context of global beauty contests). However, facts indicate otherwise: the government did, in fact, play a major role. For one, the IT industry owes much to the talent base created by the government and the policies that expanded it. The initial impetus came from high-quality engineering institutes (including from the IITs). Later, permitting private engineering institutions resulted in a huge supply of talent. Entrepreneurial initiatives also created a very large network of training institutions (NIIT and Aptech being amongst the early and better-known ones) that provided short courses in computers. These further expanded the talent base. Also, experienced and top-notch expertise existed in the large public sector (government) companies and R&D laboratories. Some of this, too, was available to fuel the takeoff of the IT industry.

While this foundation was important for growth, three other factors were vital contributors as well. The first was the reduction in import duty on software, from a peak of 150 per cent, to zero. This gave a big boost to the use of computers by different sectors of the economy, thereby providing the IT industry with substantial business opportunities. The second was the zero tax on export of software from export processing zones, which acted as a big incentive. However, this push for exports may not have succeeded without the third factor—the creation of the Software Technology Parks of India (STPIs). There were two innovations here: first, it was decided that being government organizations, STPIs could provide communication links that were, till then, the monopoly of a government department (at the time, not particularly known for efficiency, or the quality and reliability of links). The other feature was that an STPI could, in effect, be virtual. This meant that to get a tax benefit, the company need not be constrained to being in the few export-processing zones that existed; its location could be declared an STPI.

In addition, the government worked closely with the IT industry to smoothen and facilitate operations, and helped to sort out issues with other countries (e.g., work permits or visas). This partnership—and the attendant trust—were unique; traditionally, the government and the private sector were always at loggerheads. The IT sector was probably the first in which such a close and harmonious relationship was fostered.

COALITION OF COMPETITORS

A major contributor to the creation of trust and partnership has been the IT industry association—the National Association of Software and Services Companies (NASSCOM). The fact that it represents all segments of the industry is important. Its constituents (members) include small and giant companies; MNCs and Indian companies; BPOs/BPMs, IT services, products, engineering and R&D services; and Internet platform and start-up companies. Thus, it spans the gamut of the sector and its voice reflects the consensus across the IT industry. Its rigorous research, balanced analysis, evidence-based recommendations and espousal of India's interests have helped win the trust of the government.

Apart from its relationship with the government in India, NASSCOM is a strong and influential voice in global forums and with governments abroad. Within India, its initiatives in a host of areas—talent, skilling, infrastructure, support for start-ups, domestic market development, and small- and medium-sized enterprises (SMEs), amongst others—have been of immense help to the industry. Of course, policy advocacy in India and abroad, and global trade development are its special strengths. It has successfully brought together intensely competitive companies, creating a 'coalition of competitors'[97]. It was also the initiator and integrating point for the long-term vision studies of the industry, each of which looked ahead about a decade, giving the industry a common vision as well as a roadmap. Its role in the growth of India's IT industry is well acknowledged at home and globally.

Clearly, there is a strategy that has helped the IT industry be as successful as it is. The main elements, as recounted in earlier paragraphs, may be summarized as follows:

1. Developing and nurturing human resources.
2. A clear long-term vision looking ahead a decade, with a roadmap for action (similar to the ten year plans of both the atomic energy and space programmes).
3. An industry association, with great credibility amongst all stakeholders, which created a coalition of competitors, working on elements that increased the size of the pie (the market) and basic elements that benefitted all organizations in the industry.
4. Research and studies that helped to open new geographies, and also new opportunities in existing markets.
5. Innovative, industry-friendly and supportive government policies.
6. Collaboration and partnership between the government and the industry.

7. Companies and leaders that quickly grasped emerging challenges and opportunities, and shaped appropriate strategies with speed and agility.
8. Industry leaders who, while competing intensely, worked together to increase the overall market and acted jointly on basic issues that affected all.
9. Deep commitment at an individual (company) and collective (industry) level to train and re-skill existing staff in new technologies.

GEARING UP FOR INDUSTRY 4.0

Rapid technological advances are leading to the so-called 'fourth industrial revolution', or Industry 4.0. Having missed the bus in earlier industrial revolutions, what does India need to do to make sure that it grabs the opportunity this time? What policies and strategies are required to take on the challenges of the new technologies, and put India in the lead in the new world? What lessons can be transferred to the new areas from the three technologies (nuclear, space and IT) in which India has seen success?

Recent years have seen a new challenge: the rapid digitalization of all aspects of business. The IT industry has rightly looked at this as a huge opportunity. Some see it as an equivalent of Y2K or the early years of BPO, both of which gave India a big boost. Yet, this combination of digitalization and new technologies (especially automation) entering the IT industry does pose a big challenge. A major one is that automation reduces the importance of the cost and availability of talent (two of India's advantages), while also providing consistent and higher quality output. This makes offshoring far less attractive and many customers are looking at the possibility of moving work back to their home country, rather than offshoring it to India. Further impetus comes from the present political context in many countries, particularly the US, which frowns on companies offshoring work.

Indian companies, therefore, face a triple challenge: first, to build up their own digital capabilities through locating talent and re-skilling their present workforce; second, to continue providing unique or additional value to customers; and third, as chunks of a customer company's operations get digitized, to develop enough domain expertise in the particular sector so as to be able to guide and assist the customer in the process. In addition, companies have little choice but to adopt automation and new technologies in their own work, so as to retain the competitive edge.

Indications are that Indian companies have, by and large, coped well in the new environment. Some have been more adept than others at managing the balance between adopting the new and sustaining the old (the latter being bread-and-butter for the Indian IT giants, in particular); yet, all continue to grow. The strategy of each company has varied, and that is good for the industry as a whole, since one cannot be certain as to what will work. As yet, no one has bitten the dust.

The advent of new technologies has coincided with a flowering of entrepreneurship in India. This is fortuitous because the resulting start-ups have the agility and nimbleness that the new technologies demand. Many of the new opportunities relate to what can be done through an app on a cell phone. Start-ups are, therefore, able to ride on the massive growth of cell phones as also on better capabilities, with 4G driving the rapid growth of Internet-linked smartphones.

Clearly, in the present context in India and the type of technologies involved, the private sector will have a big role—more along the lines of IT than nuclear or space technologies. The technologies involved are quite disparate—ranging from robotics, to AI and data analytics, to additive manufacturing, to genetics and bioengineering. Many classify the first two into the common category of cyber-physical systems. However, even here, there is a big difference between the (literal) nuts and bolts of a robot, and the algorithms of data analytics and self-learning AI. Given this, it is arguable as to whether a single strategy framework is appropriate. Yet, despite the wide spectrum of disciplines involved, some broad, common approaches can be suggested:

1. Develop human resources in the areas/disciplines involved and also in related areas. Since these new technologies are at a high level of sophistication, a lot of emphasis will be required on doctoral-level work and post-doctoral research. A thorough overhaul of courses and curriculum is required. Innovation and multidisciplinarity is essential. The industry must work closely with academic institutions and support them with funding where necessary.

2. Encourage and facilitate R&D. This will require

government support and funding, but the industry, too, must pitch in. Tax incentives for R&D funding could help. One approach that is worth exploring is to have a consortium of companies fund pre-competitive research (in existing academic/R&D institutions or new stand-alone facilities). The government could incentivize this through tax breaks or by offering free land for setting up facilities.

3. Establish a vision document that not only lays down long-term goals, but also the specific action required from each player—industry, academia and government. A high-level working group comprising the three could do this, just as the Task Force did for IT in 1999.

4. Support start-ups in the field of emerging technologies by encouraging incubation centres and facilitating funding. A dedicated fund, initiated by the government, could help to kick-start this. A good model is the Fund of Funds created by the government for start-ups in 2016. This ensures that decision-making regarding investments (at both levels—the fund, and the individual start-up) is done by experts (not by the government) and that the end-investment (in start-ups) is a multiple of the initial fund size.

5. Use government procurement as a tool—by giving preference to Indian companies, especially start-ups— for orders being placed by the government. This could be limited (in time and/or scope), so as to act as a growth catalyst and not become a crutch.

6. The government and industry need to work together to open up markets abroad and to acquire key technologies.

7. The industry must take lead—with government facilitation and encouragement—to integrate companies from these new-tech sectors into the relevant industry association, or (only if essential) create a new one.

8. The government and industry (generally, through the relevant industry association) must work in close partnership to ensure an ecosystem conducive to the adoption of new technologies. Joint work should also remove all barriers in this regard and encourage such adoption by user-sectors of the economy, including in the social and rural sectors.

9. Develop partnerships and collaborations with appropriate organizations (Indian and foreign; academic and corporate) to seek technologies, training and experience wherever required.

10. Prioritize applications and use, and facilitate the easy import of hardware for this, wherever necessary, rather than raising barriers till indigenous hardware is developed.

The National Institution for Transforming India, also called NITI Aayog, has already formulated a strategy/policy for AI, in consultation with experts from academia and industry.[98] Similar work must now be done in other areas of new technologies. There is also a need to see how to synergize some of them, to create a greater impact. In all this, it would be worthwhile to pick ideas that can be drawn from success in other sectors.

In earlier chapters, the focus has been primarily on how

technology affects—and may change—day-to-day activities. As new technologies are developed, it is important to assess their likely impact and, importantly, to plan how they could be used to make life easier and better for all. Development of technologies and, particularly, of the systems that they can be integrated into, must be consciously guided by this objective of contributing towards a better society. This requires an understanding of societal needs in an ever changing context, identifying obstacles to the adoption of specific technologies and applications, and optimizing the beneficial impacts. Involving social scientists from the inception and integrating them into teams that develop and deploy technology is the best way of doing this. The space programme, with its active involvement of social scientists and other non-tech professionals (including a large in-house team at a crucial juncture of development), has already shown the way. It is necessary that this successful example be emulated—with adaptations as necessary—in the new technological age.

This, and the specific points enumerated earlier, could be the broad guiding strategies and policy framework for the new-tech areas. An approach along these lines for erstwhile 'new' technologies has brought us success; it should do so again.

EPILOGUE

Fast Forward to the Future

Society has now reached a stage where technology is affecting almost all aspects of life. In fact, we have gone beyond that—it has now become an integral part of our life and the ecosystem around us. Technology is not only embedded in human activities, but is increasingly—and literally—embedded *within* us. Some humans are no longer composed only of bone, brain, flesh and blood, but also of electronics, metals and plastics. Like refurbished cars, we may soon have humans with worn-out or old parts replaced by new ones.

Technology has certainly changed how we live; it has also changed *where* we live. Air conditioning and heating have made it possible to live in extreme environments. Large cities like Dubai or Las Vegas have grown and prospered, despite being located amidst hot and inhospitable deserts; so have cities in really cold environments—Siberia, for example. In the near future, it is entirely possible that there will be human settlements underneath the sea, or in outer space—on the Moon or other planets. Technology has already wrought radical change; the future promises far greater disruptions than those in the present.

The advance of technology—particularly the pace at which it is progressing—has also given rise to deep concerns. It has certainly brought much benefit to man: faster economic growth, better education and healthcare, quicker and easier connectivity, and less drudgery, amongst other things. Yet, there are those who worry whether technology has become too dominant, taking away the human touch and leading to depersonalization. Children spend their spare time watching TV programmes or playing video games by themselves, rather than playing with other children. Young adults are glued to their cell phones— reading messages, sending pictures or busy watching videos on Netflix or YouTube.

There is little doubt that direct human-to-human interaction, which is not mediated through a machine, has decreased. Of course, technology does broaden our reach and enable us to connect with those who are far away, or reconnect with long-lost contacts. In my younger days—and possibly up to the late '80s— connecting with friends or relatives abroad was only possible by post; a phone call was rare, as international calls were difficult to make and also expensive, used only for conveying special news. Making a call just to chat was unthinkable. This continued to be so, even after the major transition to international subscriber dialling (ISD), when one could directly dial someone abroad, rather than having to book a call and going through an operator. As a result, whenever I went to the US, I would spend much time every night (lower rates at night!) calling up those friends and relatives living there who I couldn't immediately meet. Now, with the convenience and ease in making international calls, and the drop in international call rates, it is common to phone friends abroad from India itself, even for a casual conversation. Such impulsive calls are, obviously, even more frequent within the country, irrespective of which distant part of India the friend/family is in. Thus, technology has helped to increase the 'width' of our interaction, with geography no longer being a constraint. However, these are still technology-mediated, and the question remains as to whether the time spent on such remote conversations is at the cost of face-to-face interaction. One indicator is that many people living in large cities do not even know their immediate neighbours, leave alone interacting with them.

Technology has, thus, helped us to conquer the constraints of geography (or distance). It has also helped us to conquer time, in a sense, through social media: sites like Facebook and LinkedIn have reconnected us with long-lost classmates and friends. Applications like WhatsApp (or older ones like email) enable the creation of common-interest groups and connect people. I know many who have located old friends and then created an online group with them, resulting in frequent and substantial interaction. This has also happened with my batchmates from half a century ago; the consequent excitement and nostalgia across the group was palpable. These could not have happened without technology. One flip side, though, is that even people in proximity communicate with each other through a device, and not directly. In the world of business and work, too, more people (and organizations) prefer video-conferencing over travelling for a meeting.

THE RISE OF MACHINES

It seems that the future is going to be one in which technology mediates most of our interactions. In fact, some of these— probably an increasing proportion—will be human-to-machine and vice-versa, as robots are embedded with greater 'intelligence' and are able to provide more services, while becoming more 'human-like' in speech. Many fear that this will lead to depersonalization and, ultimately, to dehumanization.

Another concern with regard to technology is whether it will destroy jobs. As automation, aided by robotics and AI, becomes more prevalent, will it make many jobs redundant? Today, automation ensures higher quality, easier adaptability, consistent product or service, greater throughput, and transparency. However, much of these come at a high cost, making automation economically attractive only where quality or speed is vital and cost is secondary. As costs reduce, this will change, and it will make economic sense to automate more tasks. The impact of this could be substantial and studies have identified many jobs that will be taken over by machines in the near future.[99] At the same time, there may be many new jobs—some that include altogether new functions. This would, of course, require new skills for, or the re-skilling of, the existing workforce and the introduction of new courses in educational institutions. Will the number of new jobs exceed those that become redundant, ensuring a net gain in jobs? Even if this is so, who will gain from these new jobs and who will get the short end of the stick? What will this mean for a country like India, which already has a large backlog of unemployed and underemployed people, in addition to a demographic surge of young people?

A bigger worry voiced by some eminent people is that technology literally 'takes over'; like in some science fiction scenarios, AI and robotics help machines become masters, and humans are relegated to slavery. With rapid advances in machine learning, through which machines are able to crunch massive databases and correlate specific combinations of items with outcomes, they have become evermore intelligent. Learning from errors and mistakes enables them to continuously improve their abilities and modify the algorithm they use. A large (practically unlimited) memory, the ability to process huge amounts of data at a phenomenal speed and continuous learning—these features make such machines 'super students' and very quick learners. In many functions, these capabilities grow with the size of the available database. The IoT provides this in abundance through sensors that automatically collect data and transmit it to a computer. As these devices proliferate, the amount of data available will grow exponentially. Analysis of this—and the predictions made from it—will, with machine learning, become evermore accurate.

These developments have already made machines superior to humans for certain tasks, and the list is one that is growing rapidly. Robots taking over may, on the face of it, seem fanciful and far-fetched; yet, logic indicates that it is not impossible. To start with, it may be that instead of a standard, manual (human) override for certain automatic functions, we have an automatic (machine) override for some human decisions. In a car, for example, a computer may be able to assess all parameters and make a better and quicker decision to ensure more safety in a crisis than a human. The computer will, at super speed, be able to calculate the velocity of the incoming obstacle (a sudden object, an approaching car, or a pedestrian) and—based on millions of past cases that it will draw on—model the likely behaviour of the pedestrian or the driver in the other car, and will also instantly work out the best and safest evasive action. Thus, human error— so common in these events—will be eliminated. In situations like this, an automatic override provision transferring control to a computer, may be eminently sensible. As machines get more intelligent, will not such cases proliferate to the point that most major decisions will be made by machines? Does this not mean that machines will take over?

Technology poses another challenge: it has created an existential threat to humans. There are two major possibilities that hold our future hostage. The first is a massive nuclear war that will kill millions and result in radioactivity that will threaten the survivors. The second is industrialization-induced climate change, which will ultimately make the Earth uninhabitable. Both these—while triggered by technology—are actually caused by human shortsightedness. In both cases, the control button is yet in human hands (though, in the case of climate change, it may become irreversible after a point). Therefore, it is humans—and not technology—that must be held responsible.

These are the mega-scale changes that technology may bring about; but there are also many ways in which technology plays a positive role in our day-to-day life. It is now ubiquitous and has suffused practically everything that we do. Earlier chapters have addressed only some of the areas and changes; yet, they are indicative of how deeply technology has affected many aspects of our lives, including our behaviours and attitudes. All indications are that its impact will grow, and change will be even faster.

DIGITAL DETOX

Could all this lead to a reaction? Is it possible that not just a fringe group, but a large part of the population will become neo-Luddites*, eschewing the use of technology? Many years ago, I visited one such community (the Amish) near Philadelphia in the US. They believe in simple living and are reluctant to adopt the conveniences of modern times. In many ways, they yet live in the eighteenth century, with no cars and no TV. I wonder, though, how these modern-day Luddites really get by without the use of any machine. Is it possible that we will see many such groups proliferate, trying to live the simple life of times long gone?

It is true that many people feel that technology is now too intrusive. It is not just the CCTV cameras that record your

---

*Neo-Luddism opposes many forms of modern technology. In the nineteenth century, some textile workers in England, known as the Luddites, destroyed weaving machinery for fear that these machines would make their jobs redundant.

movement in public spaces and in offices, but also mobile phones, which constantly track you and keep you connected. Unwelcome or untimely calls intrude on your private moments and societal expectation is that you respond immediately. The same is true of messages and email. It is good to feel connected and be able to reach out—whether for work or personal reasons—and get an instantaneous response. Yet, at times, most people do resent this 'electronic leash', which limits their freedom. Experts recommend a 'no-connect', device-free time as a way of reducing stress. It is already fashionable to go on a no-connectivity vacation; soon, this may even extend to a no-gadgets one—a true 'getting away from it all', like a desert-island break. Friends who go on mountaineering expeditions or trek to remote areas, and are cut off from the rest of the world for a week or two, living in rudimentary tents amidst nature, tell me that while they miss the connectivity, they undoubtedly enjoy the break. It enables a 'digital detox', and this may well become an integral part of our future living.

As tempting as such 'escapes' may seem to many—especially those who live in the high-pressure world of competitive corporates—the fact is that most people need and value the connectivity and facilities offered by modern technology. Life without various gadgets is unthinkable; living without these necessities may be a welcome change, but only for a few days.

This is somewhat similar to what I saw in Jerusalem a few years ago. Orthodox Jews observe Shabbat every week (from Friday evening to Saturday evening). The ultra-orthodox amongst them do not actively use any gadgets during that time, which even includes not switching on electrical appliances. As a result, once, during Shabbat, the hotel in which I was staying, had no toasters available for use during the buffet breakfast! The truly orthodox apparently do not switch on any electrical appliances at home either—so, no TV or microwave! However, it seems many would like to follow the letter of the religious requirement, but not necessarily the spirit. Here, technology comes to their rescue—there are pre-programmable TV sets that automatically turn on the TV at the right time so that they do not miss their favourite show! I discovered that there was a similar explanation for the intriguing fact of the lift in the hotel stopping at every floor. It was programmed to do so, to enable guests observing Shabbat to get to their rooms without having to press any buttons themselves (presumably absolving them of the irreligiosity of actively using a technology). As elsewhere, technology helps to provide unique solutions!

TIME FOR A MULTI-STAKEHOLDER APPROACH

Will the future of humanity be one in which technology is the overwhelming theme? Will art, philosophy or literature have a role at all? Indications are that technology will continue to dominate, even in the future. However, it now seems to be entering an era when creativity and innovation, and the understanding of human psychology and sociology, are becoming important. In economics, cold and logical mathematical models are no longer sufficient; behavioural economics (which takes note of the quirks of human behaviour) has become essential to understand the economy and to build predictive models. In a similar manner, as machines move to more sophisticated levels of AI, programming them to have an understanding of human behaviour will be essential. Further development of AI will have to include social science experts, from fields like sociology and psychology, as part of a multidisciplinary team.

Innovation is already a major driver of new developments. This, too, best comes from cross-disciplinary fertilization and multidisciplinary teams. Apart from social scientists, such teams would benefit from the different perspectives brought in by those from other fields of humanities, as well as the arts, theatre and the like. The interaction of such people with technologists can often stimulate innovations amongst the latter. Exceptionally, one may find a person who combines these roles within himself/ herself. One such rare person was Leonardo da Vinci, an Italian polymath, who epitomized the spirit of the Renaissance. He was, at once, a painter, a sculptor, an architect, a scientist, a mathematician, a musician and more. Little wonder, then, that he was both creative (as the painter of 'Mona Lisa' and 'The Last Supper') and an inventor par excellence (conceptualizing the early prototypes of the parachute, tank and helicopter). It seems that many of his innovative ideas were a result of his artistic bent of mind, rather than on the basis of pure logic. It is more than likely, as noted earlier, that even in times to come, machines will not be able to imitate the intuition or creative leaps that originate in the human mind and such flights of fancy are more likely to be taken by creative thinkers, rather than those whose thought processes are purely logic-based. It would seem, then, that while technology may shape the future, technology itself will be shaped as much by social scientists and artists, as by engineers.

As people worry more about the relationship between intelligent machines and humans, ethics and philosophy also will have a role to play. This is especially important in the area of bioengineering and gene editing. Custom-made babies, superhumans, immortality, the creation of new species—all these developments, spawned by the advances of science and technology, will require ethical choices and philosophical considerations to be made.

Most of the issues discussed here are not of a distant future; they are more than likely to be the reality that is confronting us in the here and now. The global community has not yet set up mechanisms to handle these vital issues. In some areas— for example, outer space and the seas—there are global treaties and conventions. We now have such international agreements in relation to the environment too (for example, the Paris Accord, signed by 193 countries). However, there is yet no equivalent for cyberspace, though some aspects of Internet governance and technological standards are now handled by a global body— the Internet Corporation for Assigned Names and Numbers (ICANN). In the case of AI/robotics and bio-genetics, there are not even preliminary intergovernmental discussions towards a global accord. It may well be time for initiating a global dialogue in these areas, which could evolve towards formal international covenants. Learning from the discussions pertaining to the Internet and taking account of the fact that governments are no longer the sole players in many areas of technology, such a dialogue must also involve the private industry, civil society and academia. This multi-stakeholder approach recognizes the reality of today's context, and the need to get inputs and a 'buy in' from across the spectrum of society.

SHAPING THE FUTURE

These macro-issues will determine our future , which, hopefully, will be one that is shaped and decided by humans, and not by technological determinism. Meanwhile, at the level of the common man and his/her humdrum daily life, technology continues to make a very big—and growing—impact. What will the future look like in terms of the routine of day-to-day living? Earlier chapters have indicated where we might be heading in some specific fields. However, in a tech-driven world, trends and predictions are meaningless. Major advances come through disruptions and bring radical changes. Examples include aeroplanes, electricity and the Internet, amongst many others. These affect not only how we do business or how we live, but also our relationship with others and the structure of society. It seems more than likely that the concept of nation states as sovereign, geographically defined entities will die. Will it be replaced by 'virtual countries' with no geographical boundaries? Will the movement of people across today's national boundaries be free? Will the family, as a social entity, disappear? While the concept of nation states is comparatively new, dating back to just a couple of centuries, the concept of family as the basic building block of communities is millennia-old. These changes are, therefore, going to be difficult.

If any or all of these happen, there will be need for new organizing principles or structures for families, industries and countries. Evolution may be faster than we think, and all of us will have to learn to adapt to the new structures that will emerge. Thus, the near future is about adopting and adapting to not only new technologies but also to new structures and relationships. Almost certainly, the latter will be more challenging than the former.

The changes ahead—some visible and many yet unknown—seem not only daunting but also challenging. They provide an exceptional opportunity to shape the future of all humans. It is, indeed, a chance to create a sustainable, happy and healthy future—one in which we live in harmony with our ecosystem and create a society that is equitable, just and compassionate. In doing this, technology can be the biggest aid—a tool that can help shape a fruitful daily life for each human, even as it assures a great collective future for humanity.

Will we face up to the challenges and fully utilize the opportunities that they present? Can we shape the new technologies—including those that we do not even know of today—into systems that work with, and for, all human beings? This will require people everywhere, across national boundaries and transgressing all divisions, to work together towards a common goal and a new global compact. The glimpses provided here—of technology in our daily life and where it may be headed—provide a context to the issues raised. As we grapple with day-to-day concerns and worry about mega matters, this book seeks to provide fodder for debate and discussion, and hopefully facilitate a consensus on what we could do collectively.

ENDNOTES

1. Gibbs, S. (2015, July 27). Musk, Wozniak and Hawking urge ban on warfare AI and autonomous weapons. Retrieved from https://www.theguardian.com/technology/2015/jul/27/musk-wozniak-hawking-ban-ai-autonomous-weapons

2. Anand, N. (2018, July 23). Indian ATMs could run dry—yes, again. Retrieved from https://qz.com/india/1333660/new-rs100-notes-could-leave-indias-atmscashless-again/

3. Wright, C. (2017, November). How Paytm went big on Indian demonetization. Retrieved from https://www.euromoney.com/article/b15ts6qpxvj51d/how-paytm-went-big-on-indian-demonetization

4. Press Trust of India. (2016, April 05). RBI's Vision 2018 for 'less cash and more digital' society on the cards. Retrieved from https://www.livemint.com/Money/9xuhzazKD20jdabvP8363O/RBIs-Vision-2018-for-less-cash-and-moredigital-society-b.html

5. Crowdfunding India. (n.d.). Let's make a 8km canal for 700 drought-hit farmers in 15 days by Suryoday Parivar. Retrieved from https://www.fueladream.com/home/campaign/216

6. Nekaj, E.L. (2016, August 18). India's Top 10 Crowdfunding Platforms. Retrieved from https://crowdsourcingweek.com/blog/indias-top-tencrowdfunding-platforms/

7. Thaver, M. (2018, August 20). What Cosmos Bank's ₹94 crore online fraud says of bank security. Retrieved from https://indianexpress.com/article/explained/what-cosmos-banks-rs-94-crore-online-fraud-says-of-bank-security-5314794/

8. Business Standard. (2018). 'Budget 2018: Govt to extend Modi's flagship PMJDY scheme, double overdraft'. Retrieved from https://www.business-standard.com/budget/article/budget-2018-govt-to-extend-modi-s-flagship-pmjdy-schemedouble-overdraft-118012300395_1.html

9. Kim, C. (2018, January 11). South Korea plans to ban cryptocurrency trading, rattles market. Retrieved from https://www.reuters.com/article/us-southkoreabitcoin/south-korea-plans-to-ban-cryptocurrency-trading-rattles-market-idUSKBN1F002B

10. Nakamoto, S. (2008). Bitcoin: A peer-to-peer electronic cash system.

11. Dua, R. (2018). Punjab, HP among top 5 states with long life span. Retrieved from https://timesofindia.indiatimes.com/city/chandigarh/punjab-hp-amongtop-5-states-with-long-life-span/articleshow/58799262.cms

12. United Nations Population Fund. (2017). Caring for Our Elders: Early Responses. Retrieved from https://india.unfpa.org/sites/default/files/pub-pdf/India%20Ageing%20Report%20-%202017%20%28Final%20Version%29.pdf

13. Press Trust of India. (2018). India's elderly population will cross 340 million by 2050. Retrieved from https://timesofindia.indiatimes.com/india/indias-elderly-population-willcross-340-million-by-2050/articleshow/65351671.cms

14. Sinha, K. (2014). Bionic hand allows amputee to feel again. Retrieved from https://timesofindia.indiatimes.com/home/science/Bionic-hand-allowsamputee-to-feel-again/articleshow/29935517.cms

15. Hignett, K. (2018). Bionic hand with a sense of touch used for

first time in real world. Retrieved from https://www.newsweek. com/bionic-hand-portabletouch-770395

16. Cookson, C. (2018). Breakthrough over growing human organs in animals. Retrieved from https://www.ft.com/content/1eff740c-148b-11e8-9e9c-25c81476 1640

17. Associated Press. (2017). Scientists make first ever attempt at gene editing inside the body. Retrieved from https://www.theguardian. com/science/2017/nov/15/scientists-make-first-ever-attempt-at-gene-editing-inside-the-body

18. The Economist. (2018). Using thought to control machines. Retrieved from https://www.economist.com/leaders/2018/01/04/using-thought-to-controlmachines

19. Sreedhar, N. (2018). Nissan brain-to-vehicle technology: What if cars could read your mind?. Retrieved from https://www. livemint.com/Leisure/tYA6haBUYX2exvtUJxq2eI/Nissan-braintovehicle-technology-What-if-carscould-read.html

20. Spinks, R. (2016, June 13). Meet the French neurosurgeon who accidentally invented the "brain pacemaker". Retrieved from https://qz.com/704522/meet-the-french-neurosurgeon-who-accidentally-invented-the-brain-pacemaker/

21. Press Information Bureau. (2018, January 12). India registers significant decline in under five child mortality rate; Rate of decline has doubled over last year. Retrieved from http://pib. nic.in/newsite/PrintRelease.aspx?relid=175583

22. Press Trust of India. (2018, January 12). India records less than a million under-five child deaths for first time in 5 years, WHO lauds effort. Retrieved from https://www.hindustantimes. com/india-news/india-records-less-than-amillion-under-five-child-deaths-for-first-time-in-5-years-who-lauds-effort/story-8yFTUHgNmgmmyvyjgiv1nM.html

23. Ministry of Human Resource Development. (2018). Educational Statistics at a Glance 2018. Retrieved from http://mhrd.gov.in/

sites/upload_files/mhrd/files/statistics/ESAG-2018.pdf

24. Pratham. (2017, January 18). Annual Status of Education Report (Rural) 2016. Retrieved from http://img.asercentre.org/docs/Publications/ASER Reports/ASER 2016/aser_2016.pdf

25. Pratham. (2018, January 16). Annual Status of Education Report (Rural) 2017. Retrieved from http://img.asercentre.org/docs/Publications/ASER%20Reports/ASER%202017/aser2017fullreportfinal.pdf

26. Kingdon, G. G. (2017, March). The Private Schooling Phenomenon in India: A Review. Retrieved from http://ftp.iza.org/dp10612.pdf

27. Saha, D. (2017, April). In 5 years, private schools gain 17 mn students, govt schools lose 13 mn. Retrieved from https://www.moneylife.in/article/in-5-yearsprivate-schools-gain-17-mn-students-govt-schools-lose-13-mn/50275.html

28. Vignesh, J., & Chanchani, M. (2018). Here's how edutech startup Byju's quietly turned into a unicorn company. Retrieved from https://economictimes.indiatimes.com/small-biz/startups/newsbuzz/edutech-startup-byjus-quietlyturns-a-unicorn-company/articleshow/63293773.cms

29. Sen, A., & Verma, S. (2018, September 03). Byju's set to raise funds at over $2 billion valuation. Retrieved from https://www.livemint.com/Companies/vj2jKqqq9dv2qWdtoN6pmN/Byjus-set-to-raise-funds-at-over-2-billionvaluation.html

30. Sarkar, J. (2018, August 26). Binge-watchers don't need data dieting any more. Retrieved from https://timesofindia.indiatimes.com/business/india-business/binge-watchers-dont-need-data-dieting-any-more/articleshow/65546992.cms

31. Ibid.

32. Gregoire, C. (2015, December 04). Netflix Binging Could Have One Major Health Consequence. Retrieved from https://www.huffingtonpost.in/entry/toomuch-tv-cognitive-decline_us_565f50c7e4b079b2818cf25d

33. Ibid.
34. Nerurkar, S. (2017). Obesity: India's big problem. Retrieved from https://www.livemint.com/Leisure/bYDJTJjcH3Ddgz1u HaqkQO/Obesity-Indias-bigproblem.html
35. Press Trust of India. (2018). India becomes largest consumer of mobile data, ranks 109th globally in mobile download speeds. Retrieved from https://www.dnaindia.com/business/ report-india-becomes-largest-consumer-of-mobile-data-ranks-109thglobally-in-mobile-download-speeds-2597890
36. Doval, P. (2018). Indians gorging on mobile data, usage goes up 15 times in 3 yrs. Retrieved from https://timesofindia.indiatimes. com/business/indiabusiness/indians-gorging-on-mobile-data-usage-goes-up-15-times-in-3-yrs/articleshow/64432913.cms
37. Telecom Regulatory Authority of India. (2018, September 18). Press Release. Retrieved from https://trai.gov.in/sites/default/ files/PRNo98Eng18092018.pdf
38. Press Trust of India. (2018). New India & Digital India? What Google says about number of people getting connected to internet in India. Retrieved from https://www.financialexpress.com/ industry/technology/new-india-digital-india-what-googlesays-about-number-of-people-getting-to-connected-internet-in-india/1056827/
39. Reuters (2016, August 05). Gangrape videos on sale in UP amid rise in crimes against women. Retrieved from https://www. indiatoday.in/india/story/gangrapevideos-on-sale-in-up-amid-rise-in-crimes-against-women-333409-2016-08-05
40. Times News Network. (2018, July). Child-lifter rumour claims 1 more. Retrieved from https://timesofindia.indiatimes.com/india/ child-lifter-rumour-claims-1-more/articleshow/64992915.cms
41. Press Trust of India. (2018, July). WhatsApp to cap message forwarding to 5 chats in India to curb fake news circulation. Retrieved https://timesofindia.indiatimes.com/business/

indiabusiness/whatsapp-to-cap-message-forwarding-to-5-chats-in-india-to-curb-fakenews-circulation/articleshow/65063622.cms

42. IndiaSpend. (2018, July). Child-lifting rumours caused 69 mob attacks, 33 deaths in last 18 months. Retrieved from https://www.business-standard.com/article/current-affairs/69-mob-attacks-on-child-lifting-rumours-since-jan-17-only-onebefore-that-118070900081_1.html

43. ETech. (2017, February). Technology sector is the second largest employer of women: Nasscom. Retrieved from https://tech.economictimes.indiatimes.com/news/internet/technology-sector-is-the-second-largest-employer-of-womennasscoms-report/57375623

44. Arora, K. (2017, November 28). Now, an app to thwart 'risky selfies'. Retrieved from https://timesofindia.indiatimes.com/india/now-an-appto-thwart-risky-selfies/articleshow/61827539.cms

45. Yadav, S. (2018, May). Passport verification can happen in 2 days, cops to visit with tabs. Retrieved from https://timesofindia.indiatimes.com/city/gurgaon/passport-verification-can-happen-in-2-days-cops-to-visit-with-tabs/articleshow/63979428.cms

46. Business Line. (2017, October). India makes it to Top 100 in 'ease of doing business'. Retrieved from https://www.thehindubusinessline.com/economy/policy/india-makes-it-to-top-100-in-ease-of-doing-business/article9935450.ece

47. Statista. (2018). Number of monthly active Facebook users worldwide as of 2nd quarter 2018. Retrieved from https://www.statista.com/statistics/264810/number-of-monthly-active-facebook-users-worldwide/

48. Statista. (2018). Leading countries based on number of Facebook users as of July 2018. Retrieved from https://www.statista.com/statistics/268136/top-15-countries-based-on-number-of-facebook-users/

49. FE Online. (February 2018). WhatsApp now has 1.5 billion monthly active users, 200 million users in India. Retrieved from https://www.financialexpress.com/industry/technology/whatsapp-now-has-1-5-billion-monthly-active-users-200-million-users-in-india/1044468/

50. Statista. (2018). Number of monthly active Twitter users worldwide from 1st quarter 2010 to 2nd quarter 2018. Retrieved from https://www.statista.com/statistics/282087/number-of-monthly-active-twitter-users/

51. Statista. (2018). Leading countries based on number of Twitter users as of April 2018. Retrieved from https://www.statista.com/statistics/242606/number-ofactive-twitter-users-in-selected-countries/

52. Greenwald, G. (2013, May 04). Are all telephone calls recorded and accessible to the US government?. Retrieved from https://www.theguardian.com/commentisfree/2013/may/04/telephone-calls-recorded-fbi-boston

53. Sherwell, P. (2013, October 27). Barack Obama 'approved tapping Angela Merkel's phone 3 years ago'. Retrieved from https://www.telegraph.co.uk/news/worldnews/europe/germany/10407282/Barack-Obama-approved-tapping-Angela-Merkels-phone-3-years-ago.html

54. Press Trust of India. (2018, May 13). India witnessed highest number of Internet shutdowns in 2017-18: UNESCO report. Retrieved from https://timesofindia.indiatimes.com/business/india-business/india-witnessed-highest-number-ofinternet-shutdowns-in-2017-18-unesco-report/articleshow/64150624.cms

55. From Clausewitz's book, *On War*, written between 1816 and 1830, and published in 1832.

56. Nye Jr., J.S. (1990). *Bound To Lead: The Changing Nature of American Power.*

57. Ibid. (2004). *Soft Power: The Means To Success in World Politics.*
58. Rao, R. (September 2016). Our diaspora has a lot to offer. Retrieved from https://www.thehindubusinessline.com/opinion/our-diaspora-has-a-lot-tooffer/article21678056.ece1
59. Rangaswami, A. (2014, December 19). Brand India shines: Tata is UK's biggest industrial employer. Retrieved from https://www.firstpost.com/business/thanktata-brand-india-takes-on-more-shine-in-uk-84141.html
60. Canton, N. (2017, August 16). We're beneficiaries of reverse colonialism: Boris. Retrieved from https://timesofindia.indiatimes.com/world/uk/were-beneficiaries-of-reverse-colonialism-boris/articleshow/60092530.cms
61. Press Trust of India. (2017, August 13). Russia's romance with Raj Kapoor lives on, 29 years after his death. Retrieved from https://www.thehindu.com/entertainment/movies/russias-romance-with-raj-kapoor-lives-on-29-years-after-his-death/article19485846.ece
62. Indo-Asian News Service. (2008, February 10). Shah Rukh Khan as popular as Pope: German media. Retrieved from https://www.dnaindia.com/entertainment/report-shahrukh-khan-as-popular-as-pope-german-media-1150157
63. Aiyar, P. (2017, August 12). Meet Japan's dancing Maharaja. Retrieved from https://www.thehindu.com/news/international/meet-japans-dancingmaharaja/article19481115.ece
64. Satija, G. (2018, June 14). Aamir Khan Is Officially The Most Famous International Star In China, People Love & Adore Him. Retrieved from https://www.indiatimes.com/entertainment/celebs/aamir-khan-is-officially-the-mostfamous-international-star-in-china-people-love-adore-him-347393.html
65. Shaikh, Z. (2018, July 12). After Dangal earns ₹1300 crore, Chinese experts want India to fill China's Hollywood void. Retrieved from https://indianexpress.com/article/entertainment/bollywood/

china-india-bollywood-movies-5256146/

66. Retrieved from: https://web.archive.org/web/20180219091024/
english.entgroup.cn/boxoffice/cn/Default.aspx?week=989

67. Retrieved from: https://web.archive.org/web/20180402230358/
english.entgroup.cn/boxoffice/cn/Default.aspx?week=995

68. Press Trust of India. (2017, November 27). China to surpass
Hollywood and Bollywood, aims to become world's largest
film market by 2020. Retrieved from https://economictimes.
indiatimes.com/magazines/panache/china-to-surpasshollywood-
and-bollywood-aims-to-become-worlds-largest-film-
marketby-2020/articleshow/61821928.cms

69. Reporting Balkans. (2016). Remembering the 1999 NATO
Bombing of Radio Television Serbia. Retrieved from https://
reportingbalkans.com/rememberingthe-1999-nato-bombing-of-
radio-television-serbia/

70. Amnesty International. (2000, June). Amnesty accuses Nato
of war crimes. Retrieved from https://www.theguardian.com/
world/2000/jun/07/balkans1

71. Ristic, M. (2015, February) Serbia Honours Chomsky for
Criticising NATO Bombing. Retrieved from http://www.
balkaninsight.com/en/article/serbiahonours-chomsky-for-nato-
comments

72. Bagchi, I. (2017). Three Warfares: China's ace weapon. Retrieved
from https://www.pressreader.com/india/the-times-of-india-new-
delhi-edition/20170813/281861528606698

73. Kimball, D. (2018, June). Timeline of Syrian Chemical Weapons
Activity, 2012-2018. Retrieved from https://www.armscontrol.
org/factsheets/Timelineof-Syrian-Chemical-Weapons-Activity

74. Riedel, S. (2004, October). Biological warfare and bioterrorism: a
historical review. In Baylor University Medical Center Proceedings
(Vol. 17, No. 4, pp. 400-406). Taylor & Francis.

75. Listner, M. (2016, February 19). Op-Ed | The continued debate

about antisatellite weapons, nine years after China's test. Retrieved from https://spacenews.com/op-ed-the-continued-debate-about-anti-satellite-weapons-nine-years-afterchinas-test/

76. Al-Heeti, A. (2018, July 31). Trump's Space Force is real, and here are the reported details. Retrieved from https://www.cnet.com/news/trumps-space-force-is-realand-here-are-the-reported-details/

77. Phys.org. (2016, October). Non-state actor likely behind US cyber attack: Clapper. Retrieved from https://phys.org/news/2016-10-non-state-actor-cyberclapper.html

78. O'Flaherty, K. (2018, May 03). Cyber Warfare: The Threat From Nation States. Retrieved from https://www.forbes.com/sites/kateoflahertyuk/2018/05/03/cyber-warfare-the-threat-from-nation-states/#1fd123281c78

79. Holloway, M. (2015). Stuxnet Worm Attack on Iranian Nuclear Facilities. Retrieved from http://large.stanford.edu/courses/2015/ph241/holloway1/

80. McGuinness, D. (2017, April 27). How a cyber attack transformed Estonia. Retrieved from https://www.bbc.com/news/39655415

81. Swaine, J. (2008, August 11). Georgia: Russia 'conducting cyberwar'. Retrieved from https://www.telegraph.co.uk/news/worldnews/europe/georgia/2539157/Georgia-Russia-conducting-cyber-war.html

82. Chapman, B. (2018, August 01). What is universal basic income and how would it work in practice? Retrieved from https://www.independent.co.uk/news/uk/politics/universal-basic-income-work-john-mcdonnell-shadowchancellor-a8472031.html

83. Press Information Bureau (2017). Economic Survey: Universal Basic Income (UBI) Scheme an alternative to plethora of State subsidies for poverty alleviation; JAM and Center State cost sharing prerequisite for a successful UBI. Retrieved from http://pib.nic.in/newsite/PrintRelease.aspx?relid=157804

84. Spudis, P. (2018, May). America's Return to the Moon: A Foothold, Not Just Footprints. Retrieved from https://www.airspacemag.com/daily-planet/americas-return-moon-foothold-not-just-footprints-180969180/

85. Knapton, S. (2015, October). Nasa planning 'Earth Independent' Mars colony by 2030s. Retrieved from https://www.telegraph.co.uk/science/2016/03/14/nasa-planning-earth-independent-mars-colony-by-2030s/

86. Mudur, G.S. (2011). Cave hope for moon house. Retrieved from https://www.telegraphindia.com/1110224/jsp/frontpage/story_13628589.jsp

87. The New York Times. (1923). Climbing Mount Everest Is Work for Supermen. Retrieved from http://graphics8.nytimes.com/packages/pdf/arts/mallory1923.pdf

88. Donahue, M.C. (2017, November 09). Dino-Killing Asteroid Hit Just the Right Spot to Trigger Extinction. Retrieved from https://news.nationalgeographic.com/2017/11/dinosaurs-extinction-asteroid-chicxulub-soot-earth-science/

89. European Geosciences Union. (2018, August 30). Deadline for climate action: Act strongly before 2035 to keep warming below 2°C. Retrieved from https://www.sciencedaily.com/releases/2018/08/180830084818.htm

90. Intergovernmental Panel on Climate Change. (1990). Climate Change: The IPCC Scientific Assessment. Retrieved from https://www.ipcc.ch/ipccreports/far/wg_I/ipcc_far_wg_I_full_report.pdf

91. Ghosh, A. (2018, August 28). As carbon dioxide levels rise, India faces big crop nutrition deficiency: Study. Retrieved from https://indianexpress.com/article/india/as-carbon-dioxide-levels-rise-india-faces-big-crop-nutrition-deficiencystudy-5329883/

92. Gillis, J., & Popovich, N. (2017, June 01). The U.S. Is the Biggest Carbon Polluter in History. It Just Walked Away From the

Paris Climate Deal. Retrieved from https://www.nytimes.com/interactive/2017/06/01/climate/us-biggestcarbon-polluter-in-history-will-it-walk-away-from-the-paris-climate-deal.html

93. Howell, E. (2018, March 28). Chandrayaan-1: India's First Mission to the Moon. Retrieved from https://www.space.com/40114-chandrayaan-1.html

94. Kremer, K. (2015). India's historic first mission to Mars celebrates one year in orbit. Retrieved from https://phys.org/news/2015-09-india-historic-missionmars-celebrates.html

95. Indian Space Research Organisation. (2016). 2015 Space Pioneer Award was presented to ISRO for Mars Orbiter Mission. Retrieved from https://www.isro.gov.in/2015-space-pioneeraward-was-presented-to-isro-mars-orbiter-mission

96. Goldenberg, S. (1998, December 30). Boom time in India as the millennium bug bites. Retrieved from https://www.theguardian.com/world/1998/dec/30/millennium.uk

97. Karnik, K. (2012). *The Coalition of Competitors: The Story of NASSCOM and the IT industry.* Collins Business.

98. Niti Aayog. (2018, June). National Strategy for Artificial Intelligence. Retrieved from http://niti.gov.in/writereaddata/files/document_publication/NationalStrategy-for-AI-Discussion-Paper.pdf

99. Scott, P. (2017, September 27). These are the jobs most at risk of automation according to Oxford University: Is yours one of them? Retrieved from https://www.telegraph.co.uk/news/2017/09/27/jobs-risk-automation-accordingoxford-university-one/

INDEX

Aadhaar, 11, 14, 82, 87, 98
Acute withdrawal symptoms, 2–5
Afghanistan, regime change in, 108
Agro-industrial complexes, 144
Artificial intelligence (AI)
  Alexa or Siri, 67
  and robotics, 125
  credit ratings, 20
  robots, powered by, 57
Aid-India Consortium, 114
Al Jazeera, 120
All India Institute of Medical Sciences
  (AIIMS), 27
Amazon Pay, 21
'anywhere and anytime' connectivity
  via mobile phones, 76
APPLE satellite, 27
Arab Spring, 108
Ariane Passenger PayLoad
  Experiment, 27
'asymmetrical warfare', 127
Automated Teller Machine (ATM),
  5–7
  mobile ATMs, 14
  obsolescence, 7
  revolution, 5

third-party providers, 6
  'white-branded, 6
Atomic Energy Commission (AEC),
  143
Atomic Energy Establishment,
  Trombay (AEET), 143

Basket case, 114
Battle of Britain, 104, 106
Beauty market, 70
Beauty parlours, proliferation of, 70
Bhabha Atomic Research Centre
  (BARC), 143
Bhabha, Homi, 142–44
Bhabha–Sarabhai strategy, 150
Bharat Interface for Money (BHIM)
  App, 8
Binge-watching, 69
Biological weapons (BW), 124
'Bionic beings' or 'cyborgs', 32
Bitcoin, 15–19. See also
  Cryptocurrencies
Blended learning, 49, 58
  'brick and click' approach, 49
Blockchain, 15, 17–19
Blood pressure measurement

instruments, 24
Bollywood
Aamir Khan, 118
Bajrangi Bhaijaan, 118
Dangal, 118
Raj Kapoor, 116
Rajinikanth, 117
Secret Superstar, 118
Shah Rukh Khan, 116
Bound to Lead, 109
BPO activities, 155
Brain implants, 36–37
Brain–computer interfaces (BCI), 34
Brain-to-vehicle (B2V) technology, 36
British Broadcasting Corporation (BBC), 105
Buddhism, 111, 114, 117
Business models, 21, 48, 76, 89, 92
Business process management (BPM), 77
Business process outsourcing (BPO), 153
BYJU's: The Learning App, 55

C3I (communication, command, control and intelligence), 104
Call centres, 77, 153–56
Cataclysmic collision, 135
Central Military Commission (CMC), 120
Chemical weapons (CW), 124
Child and maternal mortality, 37
China, 39, 53, 92–93, 101, 118, 120, 156
Chomsky, Noam, 119
Clausewitz, Carl von, 104
Coalition of Competitors, 158–60
Cold and logical mathematical models, 174
Colour revolutions, 108
Communication technology, 65, 104, 106–7

Comprehensive evaluation (CCE) method, 43
Computer Literacy and Studies in School (CLASS), 45
Computer tomography (CT), 23
Cộng, Việt, 106, 127
Cosmos Bank, 12. *See also* Digital fraud
Cow smuggling, rumours of, 74
Credit packages, 11
Cross-border
movement of people, 103
payment, 2
Cross-disciplinary fertilization, 175
Cryptocurrencies, 15–19, 128
Bitcoin, 16
blockchain 18–19
demand and volatility of, 18
growing concern, 18
quick and huge gains, 16
South Korea, 18
spike in, 16
USP, 16
See also Paperless currencies
Cultural evolution, 41
Currency notes, 1
Custom-made babies, 176
Cyber 'weapons', 130
Cyberattack, 20, 128–30
Cyber-physical systems, category of, 162
Cybersecurity, 87, 121
Cyberwarfare, 104, 128

Da Vinci, Leonardo, 175
Data analytics, 11, 30, 48, 51, 80, 87, 99, 125, 153, 162
Data mining, 96
Demonetization, 7–8
Department of Atomic Energy (DAE), 143
Depression, 69

Devolution and decentralization of power, 94
Dhawan, Satish, 147
Diarrhoea, prevalence of, 23
Die Hard 4.0, 129
Digital blood sugar measurement devices, 25
Digital Detox, 172–73
Digital fraud, 12
Digital payments, 7–9, 11
Digital thermometers, 24
Digital wallets, 14
Digital X-rays, 24
Direct Benefit Transfer (DBT), 14, 86
Distance Learning, 47–50, 55
District Information System of Education (DISE), 55
DNA structure, 33
Dot-Com Boom, 151–56

Ease of doing business, 91
Ease of living, 91–92
Echo chamber effect, 96
Educational broadcasting, 44
Educational TV (ETV) programmes, 44–45
E-governance, 83, 85
Electronic leash, 173
English textile workers, 172
English-medium schools, 55
ETV broadcasts, 45
Eugenics, 33
Extraterrestrial intelligence, 138, 140
Extraterrestrial settlement, 133, 134, 136

Facebook, 35, 94, 168
Faircent, 11
Fake news, 73–75, 96–97, 105
Fake stories, 74
Fiat currency, 15
Financial crisis, 13

Financial meltdown, 13
Financial technology, 14
Fourth industrial revolution, 160
Fragmentation, 73

Genomics, 33
Global Positioning System (GPS), 60–62
    rural development, 87
Global trade, 1–2, 39, 159
    growth, 1
Global village, 124
Global warming, 135
Goebbels, Paul Joseph, 105
Goebbelsian, 105
Google, 21, 29, 50, 61, 96
    Google Maps, 61
    Google Pay, 21
Graduate Management Admission Test (GMAT), 56
Graduate Record Examinations (GRE), 56
Guerrilla warfare, 127
Guerrilla warriors, 127–28
Guru-shishya model, 53, 55

Handshake protocol, 64
Hawking, Stephen, 58
Health indicators, 23, 37–38
Higher education, 42, 45, 47, 76
    ETV, 45
    privatization, 54–55
High-frequency trading (HFT), 12
High-risk loans, 13
Hinduism, 110–11
Hiroshima and Nagasaki, 124
Hitler, Adolf, 33
Hollywood, 118
Human error, 171
Humanization of machines, philosophical implications, 67

i2iFunding, 11
ICT, use in research, 51
India–Bhutan–China tri-junction, 121
Indian National Satellite System
    (INSAT), 46, 147
Indian Space Research Organisation
    (ISRO), 26–27, 133, 147
IndiaStack, 11
Indira Gandhi National Open
    University (IGNOU), 46
Industrial research, 51
Industry 4.0, 160
Intergovernmental Panel on Climate
    Change (IPCC), 135
International Consumer Electronics
    Show, 36
International schools, 54
Internet Corporation for Assigned
    Names and Numbers (ICANN),
    176
Internet of Things (IoT), 20
Internet-linked smartphones, 162
Iraq, regime change in, 108
Islamic State (IS) extremists, 120

Jan Dhan, Aadhaar and mobile
    (JAM), 14
Johnson, Boris, 114

Kapoor, Raj, 116. *See also* Bollywood
Kapoor, Rakesh, 80
Khan, Aamir, 118. *See also* Bollywood
Khomeini, Ayatollah, 93
Knowledge creation,
    concerns, 40
    economy, 39, 40
    links between economic growth, ,
    41
    speed of creation, 40
Kohli, F.C., 151
Kosovo War, 119
Kyoto agreement, 136

Kyoto Protocol (1997), 136

Land records, digitization of, 84–85
Learning, gamification of, 49
Less cash society, 8
Life-cycle management, 53
Life expectancy at birth, 22
Limelight Networks, 68
LinkedIn, 168
Loneliness, 68–69, 112
Luddites. *See* English textile workers
Luther King, Jr., Martin, 102

Machines, humanization of, 66–68
Magnetic resonance imaging (MRI),
    23
Mahajan, Pramod, 156
Mahatma Gandhi National Rural
    Employment Guarantee Scheme
    (MGNREGS), 86
Majoritarianism, 97
Mallory, George, 134
Mandela, Nelson, 102
Manpower supplementation, 152
Massive Online Open Courses
    (MOOCs), 47–49, 52
MCA21 project, 88
McDonald's, 113
Medical education, revamp of, 28
Medical services, 'uberization' of, 31
Merkel, Angela, 98
Military warfare, 121
Millennium city 'Gurugram', 90
Mobile telephony, low tariffs for, 65
Mobile wallet, 11, 15, 21, 92
Monero, 17
mPassport Police App, 90
MRI and CT scans, 24
Multinational corporations (MNCs),
    153
Multi-stakeholder Approach, 174–77
Musk, Elon, 35, 58

Nakamoto, Satoshi, 19
Narcissism, 78
Narrow-mindedness, 73
National Aeronautics and Space
  Administration (NASA), 147
National Association of Software
  and Services Companies
  (NASSCOM), 158
National e-Governance Plan (NeGP),
  85
National Electronic Funds Transfer
  (NEFT), 8
National Institute of Information
  Technology (NIIT), 53, 157
National Institution for Transforming
  India (NITI Aayog), 164
NATO, 119
Nazism, 102
Nehru, Jawaharlal, 142
Neo-Luddism, 172
Neo-Luddite, 172
Netflix, 68, 167
Neuralink, 35
Nirbhaya incident, 71, 97
No detention policy, 43–44. *See also*
  Right to Education (RTE) Act
Non-violence, principles of, 102
North Atlantic Treaty Organization
  (NATO), 119
Nuclear bombs, 143
Nuclear explosion, peaceful, 143
Nuclear programme, 143–146
Nuclear war, 134, 171
Nuclear-industrial, 144
Nye, Jr., Joseph, 108

Obesity, 69
Occasional slowdowns, 76
Online banking, 14
Oral Rehydration Solution (ORS), 42
Oral Rehydration Therapy (ORT), 23
Outer Space Treaty, 126

Outsourcing backlash, 155
Over-the-top (OTT) applications, 121

Pal, Yash, 45
Paperless currencies, 15–19
Paramedics, training of, 28
Paris Accord, 136, 176
Parkinson's disease, 36
Passport Seva Kendra (PSK), 89–90
  digitizing process, 89
  direct police verification, 90
Payment Apps, 2, 7
Payment Banks, 21
Paytm, 7–8, 21
Peer-to-peer (P2P) lending, 10–11
Persons of Indian origin (PIOs), 112
Plastic money, 7
Point of sale (POS) machines, 7
Polarization, 73, 97
Portfolio approach, 10
Pradhan Mantri Awas Yojana
  (PMAY), 87
Pradhan Mantri Jan Dhan Yojana
  (PMJDY), 14
Predictive analytics, 99
Preloaded money, 8
Preschool education, 54
Preventive technologies, 23
Primary health centre (PHC), 25
Privacy protection, 17
Privacy, infringement of, 100
Privatization, growth in, 54
Psychological warfare, 108, 120
Public health, 26–28
Public opinion, 107

Qualified medical personnel, shortage
  of, 25
Quasi-human machines, 68
Quick Response (QR) code, 3

Radio silence, 139

Radio Television of Serbia (RTS), 119
Radioactive contamination, 134
Raman effect, 132
Ransomware attackers, 17. *See also*
   Monero
Raveendran, Byju, 55
Real-Time Gross Settlement (RTGS),
   8
Real-time updating of bank accounts,
   5
Reliance Jio, 72
Remote consultations, 28
Reserve Bank of India (RBI), 6
Reverse colonialism, 114
Right to Education (RTE) Act, 43

Sarabhai, Vikram, 144, 146–47
Sarva Shiksha Abhiyan (primary
   education scheme), 43
Satellite Instructional Television
   Experiment (SITE), 44
Satyagraha, 102
Saudi Arabia, 112, 119–20
Search for extraterrestrial intelligence
   (SETI), 138
Sedentary lifestyles and unhealthy
   diets, 69
Selfies, 78–80
Silk Road, 39
Small entrepreneurs or micro-
   enterprises, loans to, 11
Social Media, 10, 38, 62, 73–75, 78,
   96–97, 99, 108, 121, 168
   analysis of, 99
   awareness, 38
   content exchange, 96
   cross-cultural understanding, 97
   data tapping, 10
   easy communication, 73
   fake news, 74–75
   online access, 73
   power of, 97

promotion of cross-cultural
   understanding, 97
'selfies', 78
WhatsApp, 62
Social skills, 69
Software-driven image processing
   techniques, 24
Software Technology Parks of India
   (STPIs), 157
Sophisticated data analytics, 10
Space Force, 126
Space programme, 127, 144–47,
   149–50, 159, 165
Space technology, 125–26, 149–51
Srikrishna, Justice, 12
Stalin, Joseph, 102
Sub-prime lending, 13
Sukarnoputri, Megawati, 111
Super students, 170
Superhumans, 33, 176
Suryoday Parivar, 9
Syrian refugee crisis, 107

Talk-back capability, 47
Tarapur Atomic Power Station, 150
Tata Consultancy Services (TCS),
   151
Tata Institute for Fundamental
   Research (TIFR), 142
Technologies of freedom, 92, 100
Technology enablement, 28–29
Technology-culture interaction, 64
Technology-enabled and customer
   friendly solutions, 90
Telecom Regulatory Authority of
   India (TRAI), 72
Telegraphy, process of, 95–96
Telemedicine experiment, 26–27
3D printing, 132
Trump, Donald, 105, 126
Twitter, 94–96

UGC Countrywide Classroom, 46
Ultra-nationalism, climate of, 98
Unified Payments Interface (UPI), 92
United Nations Convention on the
    Law of the Sea (UNCLOS),
    120–121
United Nations Educational, Scientific
    and Cultural Organization
    (UNESCO), 140
Universal Basic Income (UBI), 131
Unmanned aerial vehicles (UAVs),
    125
US–Soviet arms race, 102

Video clips, doctored, 73
Videos sharing, 72–73
Weapons of Mass Communication,
    101, 122
WhatsApp, 62, 74, 94, 168
World Bank, 114

Year 2000 (Y2K) problem, 152, 160
YouTube, 68, 167
Yudhoyono, Susilo, 111
Yugoslavia, disintegration of, 119

Zcash, 18
Zedong, Mao, 101

www.ingramcontent.com/pod-product-compliance
Lightning Source LLC
Chambersburg PA
CBHW070659190326
41458CB00046B/6784/J